PROGRAMS
AND
SYSTEMS
An Evaluation Perspective

THE EDUCATIONAL TECHNOLOGY SERIES

Edited by

Harold F. O'Neil, Jr.

U.S. Army Research Institute for
the Behavioral and Social Sciences
Alexandria, Virginia

Harold F. O'Neil, Jr. (Ed.) Learning Strategies

Harold F. O'Neil, Jr. (Ed.) Issues in Instructional Systems Development

Harold F. O'Neil, Jr. (Ed.) Procedures for Instructional Systems
Development

Harold F. O'Neil, Jr. and Charles D. Spielberger (Eds.) Cognitive and
Affective Learning Strategies

Jane Close Conoley (Ed.) Consultation in Schools: Theory, Research,
Procedures

Harold F. O'Neil, Jr. (Ed.) Computer-Based Instruction:
A State-of-the-Art Assessment

Gary D. Borich and Ron P. Jemelka. Programs and Systems:
An Evaluation Perspective

Gale H. Roid and Thomas M. Haladyna. A Technology for
Test-Item Writing

PROGRAMS
AND
SYSTEMS
An Evaluation Perspective

GARY D. BORICH
RON P. JEMELKA

*Department of Educational Psychology
and Instructional Systems Laboratory
College of Education
The University of Texas
Austin, Texas*

Academic Press 1982
A Subsidiary of Harcourt Brace Jovanovich, Publishers

New York   London
Paris   San Diego   San Francisco   São Paulo
Sydney   Tokyo   Toronto

ACADEMIC PRESS, INC.
111 Fifth Avenue, New York, New York 10003

United Kingdom Edition published by
ACADEMIC PRESS, INC. (LONDON) LTD.
24/28 Oval Road, London NW1 7DX

Library of Congress Cataloging in Publication Data

Borich, Gary D.
  Programs and systems, an evaluation perspective.

  (Educational technology series)
  Bibliography: p.
  Includes index.
  1. Evaluation research (Social action programs)
I. Jemelka, Ron P. II. Title. III. Series.
H61.B627   658.4'01   81-17539
ISBN 0-12-118620-2   AACR2

PRINTED IN THE UNITED STATES OF AMERICA

82 83 84 85   9 8 7 6 5 4 3 2 1

Contents

Preface

This book is about decisions, values, and systems. It is about how these concepts come together to create a network of ideas, activities, and responsibilities for program evaluation. Most of all, it is about the role that values play in binding together this network and in providing a standard for both planning and evaluating programs.

Although absent from many definitions of program evaluation, the word *evaluation* has its roots both literally and figuratively in the word *value*. We contend in this book that the field of program evaluation has woefully neglected this concept and that this neglect has prevented the field from establishing a much needed conceptual base and rationale independent of the technology (designs, techniques, instruments, and models) that has come to define it.

It would not be an exaggeration to say that during development of the field the role of the evaluator was viewed rather narrowly. Indeed, the evaluator's role has been associated, in practice, far more with the researcher than with the planner, designer, or developer. However, there is a feeling emerging within agencies, institutions, and the evaluation profession itself that this view of evaluation has led to many of the frustrations practitioners have experienced with evaluation. This book is an attempt to widen the perspective with which the field of evaluation can be defined. We do this by suggesting how the evaluator's role can be viewed more comprehensively to include the milieu of values within which every program functions and to which every program must in some sense be accountable. Our task in describing this broadened outlook will be to show that the evaluator can perform functions of benefit to agencies, organizations, and programs that are not now being performed, are badly needed, and are basically evaluative in character.

Although this book addresses both program planning and evaluation, it draws upon planning and evaluation principles and concepts from many fields of inquiry. These fields include general systems theory, operations research, systems engineering, and management science, as well as contemporary movements in the field of evaluation itself. Too often, it seems, professionals retreat to the comfort and familiarity of their own disciplines or subdisciplines when faced with a complex problem, ignoring the many conceptually rich ideas that may exist in adjacent and peripheral areas. The greatest stumbling block to the interdisciplinary transfer of ideas in the area of program planning and evaluation is not a dearth of good ideas, but rather too narrow a perspective on the part of program planners and evaluators to take advantage of these ideas. Thus, we have brought together in this book a variety of interdisciplinary ideas and concepts for program planning and evaluation. We have organized these ideas under the general theme of a *systems perspective*.

Although many social and behavioral scientists speak of programs, seldom do they refer to them as systems. For example, there is the prevalent assumption in our technocratic society that almost anything can be created and managed by dividing it into its functional elements and by separating it from the larger program, system, or organizational unit to which it belongs. Some planners and evaluators believe that a program can be developed in the same way—by attending to the function and performance of each of its parts. To do so, however, ignores the fact that outcomes of some parts must build upon prior outcomes in order to achieve the larger ends to which the program is contributing. This observation highlights the importance of planning and evaluating programs in relation to the larger program, system, or organizational context in which they must function, and reveals the pitfalls of any conceptualization of planning, development, and evaluation that prevents us from seeing programs as means to subsequent ends.

The Perspective of This Book

A *systems perspective of evaluation* is a fairly comprehensive yet succinct way of describing what we have to say. But we were concerned about what that term might mean to some. It has been our experience that some of the ideas and much of the writing released under the rubric of "systems," "systems approaches," and "systems thinking" has confused or alienated many professionals exposed to them.

However, our intent is to provide the reader with a *perspective*, not an approach. We will not be "selling" or "pushing" any specific models or techniques in this book. The evaluation literature is already brimming with many specific approaches to program evaluation. We believe that many of these attempts to describe the field of evaluation are only symptomatic of the value-related problems plaguing the field and contribute little to establishing a much needed conceptual rationale for evaluation. Instead of following this path, our goal has been to *synthesize* and *integrate* into meaningful constructs some new interdisciplinary ideas for the planning and evaluation of programs. We have borrowed a few good ideas from the systems literature, melded these with some theories about values, and integrated this with what we consider to be good ideas in the field of program evaluation. We believe some of the ideas from each of these sources are potentially valuable to evaluative inquiry, whereas others are probably best discarded.

Organization and Content of This Book

This book is written for a broad array of readers interested in program planning, evaluation, or systems. It is organized into nine chapters and an appendix. Chapter 1 traces the contemporary development of evaluation in the fields of education and human service and serves as an introduction to and integration of the many definitions and interpretations given to program evaluation. It concludes with an agenda of some possible new directions for evaluation which are presented in later chapters.

Chapter 2 reviews and interprets the contemporary movements described in Chapter 1 in light of a decision-oriented perspective. While pointing out that decision-oriented evaluation is the most common perspective employed in the evaluation of programs, this perspective is noted to be the most limited in the kinds of information it can provide program managers and administrators. Examples of these limitations are discussed.

One of the possible future directions for evaluation discussed in Chapter 2 is that of value-oriented program evaluation, the topic of Chapter 3. A value-oriented approach to evaluation is formulated with Dewey's (1939) theory of valuation and Cooley and Lohnes's (1976) updating of this theory.

In Chapter 4 major concepts and trends that have traditionally earmarked systems approaches are reviewed. These concepts and trends provide a backdrop for the blending of systems thinking and evaluation which is undertaken in subsequent chapters. We include in this chapter discussion of general systems theory and practical systems models for designing, developing, and evaluating programs.

Chapter 5 discusses the limitations and atheoretical nature of much of the systems literature. Some early evaluation literature is discussed to show the historical roots of systems thinking in evaluation and the compatibility of systems concepts to evaluation. We discuss and relate systems ideas to contemporary movements in information processing, organizational development, and qualitative methodology.

In Chapter 6 we relate our evaluation perspective to programs and elucidate how and why programs should be viewed as systems. Relationships are drawn to the systems concepts of earlier chapters and especially to the concept of value-oriented program evaluation. Here, decisions, values, and systems are interrelated and discussed from a single perspective.

Chapter 7 illustrates how both the systems and value perspectives can be integrated and applied to the program planning and evaluation process, taking the criminal justice system as an example of the most general level. Various programs that can be derived from this system are then used in subsequent examples to clarify, expand, and apply the notions presented in earlier chapters.

Chapter 8 sketches a framework or perspective for applying systems notions and concepts to program planning and evaluation functions. This perspective engages the evaluator in the process of "looking up" to a higher level program, system, or organizational unit of which the program can be considered a component part, to better understand the values upon which a program is based. Criteria for evaluating program merit or worth can be found in the larger context to which the program is expected to contribute. In this chapter, as in all earlier chapters, the concept of *values* is the preordinate concept with which the evaluator works.

Finally in Chapter 9 we summarize our thesis that evaluation exists because there are *decisions* to be made about *systems* based on *values*. This chapter makes the case that while values and systemic relationships constitute "new ground" for evaluators, these concepts represent evaluation's best chance for establishing a much needed theoretical base and rationale. This new base is sufficiently broad to encompass traditional evaluation methods as well as value-oriented, naturalistic, and qualitative concepts and ideas for program evaluation. We conclude this chapter with the notion that the evaluator comes to know and understand the program context through the milieu of values, needs, and conditions that comprises it.

The appendix to this book contains a brief description of a program modeling technique used in Chapters 6, 7, and 8 to describe, clarify, and communicate the milieu of values in which a program functions. It is offered as a self-instructional guide as well as a review of key ideas for those who may want a practical means of applying some of the notions in this book.

Acknowledgments

The ideas in this book have long been in the making and we have many individuals and agencies to thank for providing sounding boards and "laboratories" for implementing and modifying its contents. Foremost among these are the first author's students who, over the past several years, have provided valuable insights and criticisms of many parts of this book as part of his Seminar in Program Evaluation. In addition to his students there have been several private and public agencies and institutions with which the authors have worked during the development of these ideas. Among these are SofTech, Inc., Waltham, Massachusetts, to whom both authors would like to express a special sense of gratitude for their support during the authors' evaluation of their Structured Analysis and Design Technique. The authors would also like to thank Karen Kirkhart for acquainting them with much of the mental health evaluation literature they have cited, Larry Browning for sharing his ideas on qualitative research methodology, and Tim Ruefli for expanding their knowledge of systems theory. The authors express their appreciation to Lee Cronbach, David Krathwohl, T. A. Ryan, Bill Cooley, and Bob Stake for reviewing and commenting on previous drafts of this manuscript. Finally, the authors thank Diane Linimon for her careful and conscientious editing and Jana Schultz-Mendoza for her considerable typing and bibliographic efforts over the last 2 years.

PROGRAMS
AND
SYSTEMS
An Evaluation Perspective

1

The History
and Growth of
Program Evaluation

Traditionally, the primary purpose of evaluation has been to provide decision makers with information about the effectiveness of some program, product, or procedure. Within this perspective, evaluation has been viewed as a process in which data are obtained, analyzed, and synthesized into relevant information for decision making. Although the specific approach used and procedures employed vary from one evaluation study to another as a function of who is doing the evaluation, the context in which the evaluation is to occur, and the desires and needs of the individual or agency contracting the evaluation, most evaluation activities fit comfortably within the bounds of this definition. Despite differences in the conceptual frameworks used by practitioners, there has been basic agreement about the decision-making role of evaluation.

For example, the Phi Delta Kappa National Study Committee on Evaluation (Stufflebeam, Foley, Gephart, Guba, Hammond, Merriman, and Provus, 1971) defined educational evaluation as "the process of delineating, obtaining, and providing useful information for judging decision alternatives [p. 40]." They divided evaluation into a four part process consisting of context, input, process, and product evaluation, each with its own purpose and method. Other prominent educational evaluators (Alkin, 1969; Hammond, 1973; Metfessel and Michael, 1967; Provus, 1971; Stake, 1967) conceptualized and partitioned the process of evaluation in other ways, but generally agreed on the decision-oriented purpose of evaluation.

Program evaluation's developmental roots also run deep in the field of human service delivery, where a similar decision orientation can be found. Hagedorn, Beck, Neubert, and Werlin (1976) define evaluation as "a systematic set of data collection and analysis activities undertaken to determine the value of a program to aid management, program planning, staff training, public accountability, and promotion [p. 3]." They state that in its most basic form evaluation involves comparison of program performance against some standard, usually data from a similar program, judgment by experts, or some group's preference. This definition is representative of those found in the mental health evaluation literature (cf. Anderson and Ball, 1978; Rossi and Williams, 1972; and Weiss, 1972a) and reflects the same view of evaluation's purpose as prevails in education, that of an information feedback process to administrators, or as Attkisson, Hargreaves, Horowitz, and Sorensen (1978) put it "a support system essential to decision-making at every administrative level of the care system [p. xiv]."

Certainly, this unity of purpose is not surprising when one considers the evolutionary roots of program evaluation, be it in education or human service. There has always been a thread of practicality in the evaluation of programs. That is, there has always been a need for information about something. When the "something" has been a program, product, or procedure in education or human service, the search for and acquisition of information has most often been called evaluation. Despite disparity in the terms used across fields, all of these definitions refer to some act of supplying information appropriate to some premeditated question, the answer to which is believed to have practical utility for subsequent decision making.

The purpose of this chapter is to acquaint the reader with the evolution and growth of the practical, decision-oriented nature of evaluation. This purpose, however, is not as straightforward as it may seem. It is our belief that while the evolution of decision-oriented program evaluation makes perfect sense, this evolution has resulted in a narrow perspective that has limited program evaluation as a field of inquiry. This perspective is one in which the criteria with which a program is to be evaluated are determined with little or no input from the evaluator. It will be argued that guided by practicality, convenience, and economy, decision-oriented evaluation has evolved into an eclectic, atheoretical set of procedures to be utilized when a specific information need arises. This, we believe, has reduced the role of evaluation in planning, developing, and improving educational and human service programs.

This chapter traces the development of program evaluation as it "grew up" in the fields of education and human service. The first section of the chapter describes the social context of the development of evaluation in education, the political concerns that shaped the decision-oriented perspective in educational evaluation, and the models of educational evaluation that resulted from these forces. The second section of the chapter describes the social and political forces that resulted in a shift in emphasis in program evaluation from education to human service delivery. We will see that program evaluation has developed in response to the pressing demands of a rapidly changing social system and that attempts by evaluators to meet these pressing demands have resulted in the development of decision-oriented models of evaluation. This focus of evaluation on decision making has had two important implications for evaluation as a field. First, it has resulted in the development of evaluation methodologies in the absence of an adequate underlying conceptual framework. Second, it has limited the role of the evaluator to that of technician in the service of the decision maker. We conclude this chapter by introducing a broader and, we believe, more important role for evaluation and the implications of this role for program evaluation. We are not alone in taking this perspective with regard to the evolution of the state-of-the art of evaluation practice. Cronbach, Ambron, Dornbusch, Hess, Hornik, Phillips, Walker, and Weiner (1980) have raised many of the same issues and concerns in the appropriately titled volume, *Toward Reform of Program Evaluation.*

The Growth of Decision-Oriented Program Evaluation in the Field of Education

While many events have provided impetus to the developing field of educational evaluation from a historical perspective, several movements and trends within education have greatly shaped the field conceptually.

SOME MOVEMENTS AND TRENDS SHAPING
THE FIELD OF EDUCATIONAL EVALUATION

A major impetus to the development of decision-oriented evaluation in the field of education was the curriculum reform movement. During the 1960s, primarily, this reform movement was characterized by widespread change in the philosophy, techniques, and materials used to teach elementary and secondary school children. Most notable were initiatives that occurred in the sciences shortly after the 1957 launching of the Russian satellite, Sputnik. With Soviet competition in the sciences came impetus for the federal government to play an increasing role in the field of education, initially through the vehicle of the National Science Foundation and later through the efforts of the U.S. Office of Education and National Institute of Education. The post-Sputnik era provided the context for the development of new curricular materials in the fields of science and mathematics. These initiatives represented efforts to not only revise and update certain segments of the school curriculum but also to try new approaches to curriculum development that placed less emphasis on the role of single authors of textbooks and curricula and greater emphasis on teams of specialists brought together by public monies specifically for the purpose of infusing the school curriculum with the latest scientific advances.

Of greatest significance to the field of evaluation was the fact that with a more systematic approach to curriculum development, the previously isolated concepts of instructional development and evaluation were drawn closer together. Because of the experimental nature of much of the content and approaches used, pilot and field testing of instructional components became logical extensions of these efforts. It was in this context that many nation-wide curriculum reform projects were begun. These projects contributed significantly to the field of evaluation by employing development strategies that required testing and revising component parts of curricula, a process which later was to become known as *formative evaluation* (Scriven, 1967).

Despite the influence of the curriculum reform movement, there was still relatively little emphasis placed on the evaluation of educational programs by the mid-1960s. It was in this context that the U.S. Congress began debate on the Elementary and Secondary Education Act (ESEA) of 1965. This comprehensive and ambitious educational legislation was to make available large sums of money in the form of grants to local universities and education agencies for educational materials, development, and research. Motivated by a concern for accountability, the Congress insisted on a provision to ESEA requiring that evaluations be conducted for each project funded. Guidelines were conveyed to prospective grantees in an ESEA Title III manual published by the U.S. Office of Education requiring the applicant to describe the evaluation methods and instruments to be used in conducting the evaluation and the cost of the evaluation. Although the final version of the bill did not require the evaluation of all programs (titles) under ESEA, there was a clear mandate from those providing federal funds for education that agencies utilizing these funds be accountable for the educational programs, products, and procedures they developed and implemented. For the first time educators were *required* to devote time and resources to evaluating their own efforts.

The concept of *accountability* emerged as an outgrowth of the ESEA legislation. Federal agencies and grantees responsible for innovative ESEA programs were only the first to feel the pressure for accountability. The accountability demanded of programs also raised questions about the performance of the schools and teachers who played prominent roles in their implementation. Consequently, teaching effectiveness and the administrative accountability of schools in general often became the focus of attempts to monitor and evaluate federally funded programs. The concepts of *accountability, cost–benefit,* and *quality assurance* filtered down in spirit, if not in substance, to the local school and school teacher. Some school administrators responded by mounting special programs to increase pupil performance on nationally standardized tests, college admissions tests, the National Merit Scholars examination, and the like. Others began exploring ways to make more cost-effective decisions about the operation and management of their schools, in order to prove that increased revenues actually produced more efficient teaching and learning. School administrators embraced accountability procedures in answer to community pressures for more objectively determined and effective ways to spend school revenue and to make internal decisions that could be defended to school boards, PTAs, and citizens' groups.

It was within this context of widespread community concern about higher but apparently unproductive school expenditures that some state governments began discussing legislation requiring the evaluation of school district personnel. An example of state-enacted accountability legislation is California's Stull Act, passed in 1971, requiring that school boards in that state evaluate their educators yearly and provide recommendations for their professional development. The Stull Act gave local communities a mandate to develop procedures for evaluating school district personnel and for periodically reporting evaluation data back to the teacher and administrator in order to upgrade his or her performance. For more on this topic see Borich (1977) and Borich and Madden (1977).

RESPONSE TO THE DEMAND FOR EFFECTIVE EVALUATION

Although educators were generally positive about the explicit mandates contained in ESEA legislation and California's Stull Act, it became evident that educational administrators at the federal, state, and local levels were not prepared to implement either of these new mandates effectively. Moreover, the sudden increase in demand for capable evaluators brought about by these mandates quickly exhausted the supply. Few educators had any formal training in evaluation and often local school personnel were pressed into service as program evaluators. With the emergence of ESEA came the need for trained evaluators schooled in both the concepts and practice of program evaluation.

Another obstacle to the implementation of these mandates was the inability of local, state, and federal administrators to apply the mandates. The evaluation concepts created by educators in the preceding decade no longer seemed adequate to answer the questions that were now being asked about these programs. After reviewing the evaluation reports of a set of ESEA programs, Guba (1969) concluded:

The traditional methods of evaluation have failed educators in their attempts to assess the impact of innovations in operating systems. Indeed, for decades the evidence produced by the application of conventional evaluation procedures has contradicted the experiential evidence of the practitioner. Innovations have persisted in education not because of the supporting evidence of evaluation but despite it [p. 28].

At another point he argued:

When the evidence produced by any scientific concept or technique continually fails to affirm experiential observation and theory arising from that observation, the technique may itself appropriately be called into question [p. 30].

With the emergence of ESEA came not only a need for trained evaluators but also a need for improved evaluation methods and techniques.

Reflecting on the state of evaluation practice, the report of the Phi Delta Kappa (PDK) National Study Committee on Evaluation (Stufflebeam et al., 1971) concluded that evaluation was "seized with a great illness [p. 4]." The "symptoms" of this illness, as stated by the PDK Committee, were:

1. The Avoidance Symptom—Evaluation is perceived as a painful process that may expose a school district's problems or an individual's shortcomings. Evaluation is avoided unless absolutely necessary.
2. The Anxiety Symptom—Evaluation evokes anxiety. The educator as well as the evaluator knows how cursory, inadequate, and subject to error the evaluation process can be. The ambiguity in the evaluation process engenders anxiety in both the educator and evaluator.
3. The Immobilization Symptom—Despite federal requirements to evaluate, meaningful evaluative data on educational programs, products, and procedures are still rare. This lethargy and lack of responsiveness is symptomatic of deeper ills.
4. Lack of Theory and Guidelines Symptom—There is a lack of unified theory of evaluation. With evaluators differing among themselves about what evaluation should and should not be, the evaluator in the field is left to his own devices for conducting evaluative inquiry; there are few guidelines and obtaining advice from an evaluation "expert" is no guarantee that a technically sound evaluation report will result.

And to these were added the lack of trained personnel, the lack of knowledge about decision processes, the lack of values and criteria for judging evaluation results, the need to have different evaluation approaches for different types of audiences, and the lack of techniques and mechanisms for organizing, procuring, and reporting evaluative information. The foregoing suggests that the relatively new discipline of evaluation was indeed besieged with problems that could be conceptualized as deficiencies. These deficiencies, however, were symptoms of more fundamental ills: the lack of an adequate definition of evaluation, the lack of adequate evaluation theory, and a failure on the part of evaluators to fully appreciate the context in which evaluation occurs.

DECISION-ORIENTED EVALUATION PARADIGMS

The lack of an adequate theoretical base for the discipline of evaluation has often been cited as a factor that has stifled the development of the field and its ability to provide meaningful evaluative data to practitioners. Even more problematic, however, was the lack of consensus among evaluators as to how evaluations should be conducted. Evaluation within the field of education was arbitrarily defined in a number ways. Three decision-oriented notions of educational evaluation that predominated during the early development of the field were elucidated by Pace and Friedlander (1978) and are summarized below.

Educational psychology models. These evaluation models drew their primary impetus from the early work of Tyler (Smith and Tyler, 1942; Tyler, 1942). Generally these evaluation paradigms employed the following steps: (a) identifying objectives, (b) stating objectives in measurable behavioral terms, (c) devising and administering measurement instruments, and (d) comparing obtained results with the objectives prespecified in Step a. Common features of educational psychology models include (a) their almost exclusive application to the evaluation of curriculum and instruction, (b) the high priority they place on defining objectives, and (c) their limited focus on comparisons of "what is" with "what should be," with discrepancy information constituting the bulk of evaluation results.

Stake's early work (Stake, 1967) is representative of this genre of evaluation paradigms. Stake partitioned the learning process into three components: antecedents (conditions existing before instruction, usually the entry level skills of students), transactions (the intervening program activities), and outcomes, or program results.

To Stake, the descriptive phase of the evaluation process involved (a) examining the logical relationships that exist among intended antecedents, transactions, and outcomes, (b) determining the congruence among intended and observed antecedents, transactions, and outcomes, and (c) determining the empirical contingencies among observed antecedents, transactions, and outcomes (see Figure 1.1). Illogical contingencies constituted possible program weaknesses.

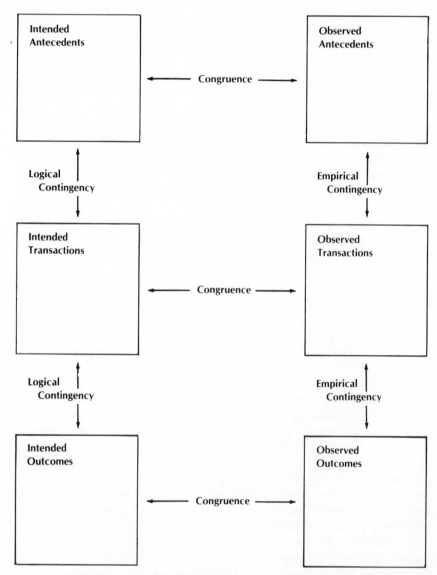

Figure 1.1. The descriptive phase of evaluation (from Stake, 1967).

Another educational psychology model commonly used during this period was that developed by Provus (1971). Provus proposed a five-stage evaluation process:

Stage 1. Documenting program description. The evaluator coaxes from program staff a comprehensive program description, including a description of all inputs, processes, and outcomes.

Stage 2. Observing field operations. The evaluator collects information as to the discrepancy between expected and actual installation. The evaluator controls and "tightens up" implementation of the program, if needed.

Stage 3. Relating component parts of the program to short-term, enabling behaviors. The evaluator collects information to show that student behavior is changing as predicted. Emphasis is on measuring enabling, or short-term outcomes, as opposed to end-of-program outcomes.

Stage 4. Relating component parts of the program to end-of-program behaviors. The evaluator collects information to show that end-of-program behaviors have been attained at some minimal level. Pre-post behavior changes and, sometimes, control versus experimental comparisons are made.

Stage 5. Comparing the experimental program with a realistic alternative. The evaluator establishes an experimental or quasi-experimental design to show that benefit is commensurate with cost.

At each stage of this process are three activities: (a) agreeing on program standards; (b) determining if a discrepancy exists between aspects of a program; and (c) using discrepancy information to identify program weaknesses. Discrepancy information at each stage leads to a decision whether to proceed to the next stage or to alter either program standards or operations. Advancement to a subsequent stage is contingent on attaining congruence between operations and standards at the previous stage. If congruence is not possible program termination is recommended, although in practice this operation is rarely chosen.

Although the evaluator's role as an information provider is not as explicit in the educational psychology models as in our next set of models, these models substantially limit the evaluator's freedom in an evaluative context (most are step-by-step procedures) and do not afford the evaluator the opportunity to examine the appropriateness of program objectives with criteria external to the program itself. (See Furst, 1964; Provus, 1971; and Tyler, 1950 for a more thorough exposition of this notion of evaluation.)

Educational decision models. Some writers in the field of evaluation have explicitly put forth the notion that the role of evaluation is to provide decision makers with information (cf. Alkin, 1969; Stufflebeam *et al.,* 1971). The following premises characterize this notion of evaluation:

1. The decision maker determines what is to be evaluated and may even choose the measures to be used.
2. The evaluator's role is that of a subservient advisor to the decision maker.
3. The "work" of evaluation consists almost entirely of information gathering and reporting.
4. The information gathered should be relevant to the decision maker's needs.
5. What information is important is dictated by the decisions to be made (Pace and Friedlander, 1978).

Alkin's (1969) definition is most representative of this school of evaluation. He views evaluation as the process of ascertaining the decision areas of concern, determining the appropriateness of information to these decision areas, and collecting and analyzing information in order to report summary data useful to decision makers.

The CIPP evaluation model is also representative of this perspective on evaluation and has achieved some notoriety among educators and evaluators. This model, developed by the Phi Delta Kappa Commission on Evaluation (Stufflebeam *et al.,* 1971), divides evaluation into four distinct strategies—Context evaluation, Input evaluation, Process evaluation, and Product evaluation, thus the acronym CIPP. Context evaluation has as its objective to specify the operational context and to identify problems underlying needs. Input evaluation is concerned with identifying and assessing system capabilities. The objective of process evaluation is to identify defects in procedural design or implementation and to document project activities. The goal of product evaluation is to relate outcome information to objectives and to context, input, and process information. If these relations are not specifiable, program weaknesses are suspected. The CIPP model allows the measurement of program outcomes both during and after program completion. At each stage outcomes are compared to stated objectives and differences between expected and actual results are reported to decision makers (see Table 1.1).

Educational science models. In this third view of educational evaluation, evaluative inquiry is conducted in much the same manner as a laboratory

Table 1.1
The CIPP Evaluation Model[a]

	Context evaluation	Input evaluation	Process evaluation	Product evaluation
Objective	To define the operational context, to identify and assess needs in the context, and to identify and delineate problems underlying the needs	To identify and assess system capabilities, available input strategies, and designs for implementing the strategies	To identify or predict, in process, defects in the procedural design or its implementation, and to maintain a record of procedural events and activities	To relate outcome information to objectives and to context, input, and process information
Method	By describing individually and in relevant perspectives the major subsystems of the context; by comparing actual and intended inputs and outputs of the subsystems; and by analyzing possible causes and discrepancies between actualities and intentions	By describing and analyzing available human and material resources, solution strategies, and procedural designs for relevance, feasibility and economy in the course of action to be taken	By monitoring the activity's potential procedural barriers and remaining alert to unanticipated ones.	By defining operationally and measuring criteria associated with the objectives; by comparing these measurements with predetermined standards, and by interpreting the outcome in terms of recorded input and process information
Relation to decision making in the change process	For deciding upon the setting to be served, the goals associated with meeting needs and the objectives associated with solving problems (i.e., for planning needed changes)	For selecting sources of support, solution strategies, and procedural designs (i.e., for programming change activities)	For implementing and refining the program design and procedure (i.e., for effecting process control)	For deciding to continue, terminate, modify or refocus a change activity, and for linking the activity to other major phases of the change process (i.e., for evolving change activities)

[a] From Stufflebeam et al. (1971). (Used with permission of Peacock Press.)

experiment. The terms *evaluation research* and *applied research* reflect this attempt to apply the experimental designs and statistical tools of classic research methodology to practical evaluation problems.

The evaluation model of Astin and Panos (1971) is illustrative of this view of educational evaluation. They proposed a causal model consisting of three components: inputs (usually student characteristics or attributes), the program (those experiences to which the students are exposed), and outcomes (certain targeted skills and abilities measured at program completion). The purpose of educational science models is to establish causal connections between these components. Often, but not always, multiple regression or linear models techniques are used to compute predicted outcome scores on the basis of various input data. Then, groups of students, subjected to various program experiences, are tested to determine whether the students in the various programs exceed, match, or fall short of what would be expected from input data alone. Program impact is inferred if outcome variance cannot be explained by input data alone.

Although educational science models à *la* Astin and Panos have not proliferated, a sorting through of evaluation studies reveals a strong reliance on the scientific method and an even heavier emphasis on the experimental designs and statistical tools of research. This result is not surprising considering that the typical evaluator is usually extensively trained in the methodology of research and often only minimally trained in those concepts unique to evaluation.

Despite obvious advantages of classical research methodology, such as experimental control over variables and the statistical power of parametric statistical techniques, there are practical considerations that limit the applicability of these procedures to educational problems. These were presented by Stufflebeam *et al.* (1971) and are updated and summarized below with some extensions and modifications.

1. *Laboratory antisepsis.* Cooley and Lohnes (1976) point out that scientific research attempts to validate the existence of cause-and-effect relationships with the ultimate goal being the development of a consistent and parsimonious theory of natural phenomena. Evaluation research, on the other hand, is concerned with means–end relationships with the ultimate goal being a rational choice between alternatives for action. Because scientific research pursues universal laws, knowledge must be obtained in a context-independent way. Experimental manipulation is used to control all confounding and extraneous variables. The evaluation of an educational program is concerned, however, with all the mitigating variables affecting some outcome. "In order to provide useful data, educational evaluation

does not need the antiseptic world of the laboratory, but the septic world of the classroom and school [Stufflebeam *et al.*, 1971, p. 22]." Laboratory research designs require conditions usually not attainable in educational contexts.

2. *Effects of intervention.* In scientific research, variables are manipulated by the experimenter to create critical comparisons of the ways variables interact. Thus, the experimenter's intents become part of the data. The evaluator, on the other hand, attempts to assess interactions in a natural rather than contrived environment. Intervention by the evaluator must be kept to a minimum and data collection must be done unobtrusively in order to avoid the confounding of evaluation results.

3. *Terminal availability of data.* Research designs typically attempt to assess the effect of some experimental treatment. The treatment is administered, then data are collected and analyzed. Data for making judgments are available only after the treatment has been administered. This precludes the use of data to refine a treatment. However, continuous refinement of an ongoing program is frequently a necessary function of evaluation.

4. *Single treatments only.* For purposes of experimental control, scientific research requires that a treatment be evaluated alone. If several treatments are operating simultaneously, their effects will confound each other. Educators, on the other hand, cannot withhold a potentially beneficial program from students because they are concurrently enrolled in other programs, activities, or classes.

5. *Effects of control variables.* Random assignment of students to programs is generally not possible in educational settings. Thus, to equate treatment groups (in order to enhance their comparability) evaluators usually match groups on selected control variables such as intelligence levels, ethnic mix, classroom size, socioeconomic status, and the like. The problem with this procedure is that criterion variables (such as measures of cognitive or affective achievement) are often correlated with these control variables. Therefore, the effects of alternative treatments tend to be obscured.

6. *Inapplicability of assumptions.* Some assumptions underlying the use of parametric statistical procedures may not be met in the usual evaluation setting, for example when distributions are severely skewed, relationships are nonlinear, or group variances are unequal.

7. *Restricted decision rules.* Conventional statistical techniques contain decision rules of the simple "go–no-go" variety. A null hypothesis may be rejected or not rejected, or treatment X may be judged better than treatment Y. Evaluators are often asked to bring their expertise to bear in more complex decision settings.

Although all exclusive distinctions between research and evaluation have been subjects of controversy, many evaluators and researchers implicitly support a broad separation between these two modes of inquiry. So sharply has the line between the educational science model and evaluation been drawn at times that some evaluators contend that the two modes of inquiry are basically incompatible and ultimately must employ different methodology (Guba and Lincoln, 1981; Hemphill, 1969). We shall have more to say about evaluators' reliance on experimental design and related statistical tools and techniques at a later point. It should suffice to say here that the problems associated with this practice go beyond those cited by Stufflebeam et al. (1971).

Summary

A major advantage of the educational psychology, educational decision, and educational science models was that they forced educators to conceptualize and clearly articulate the goals of instruction. Moreover, their emphasis on objectives provided at least implicit criteria for judging the success of a program. However, these advantages can be weighed against several significant limitations of these general approaches to evaluation. These limitations include:

1. *Focus on interim outcomes.* Since objectives must be specific to be measurable, longer term and higher order objectives intended by program developers may have been eliminated from the evaluation when they had to compete with concrete and short term objectives which were more easily defined and measured. This may have been especially true when program effects were not expected to fully accrue during the program.

2. *Emphasis on measurement.* The emphasis these models place on measurement may have obscured judgments and judgment criteria. Test scores could become entities unto themselves while the concepts behind the scores became obfuscated. With highly precise modes of quantification come the illusion that the data themselves are what is important, not the judgment criteria by which the data must be interpreted and made meaningful.

3. *Emphasis on student achievement.* These models place heavy emphasis on student achievement. Thus, a new staffing policy or instructional strategy was likely to be evaluated in terms of student achievement, and such issues as cost-effectiveness, teacher satisfaction, and student discipline were ignored.

4. *Terminal availability of data.* The data emphasized with these models, that of student performance, were available only at the end of the project when the performance of students was compared to program objectives. Thus, valuable process and instructional data were often not collected (or at least not measured as comprehensively) and the opportunity for feedback to program personnel for program modification was diminished.

5. *Limited judgment criteria.* In these models program objectives and the criteria by which they were to be evaluated were generally "givens" and the evaluator was expected to collect and interpret data only in relation to the "givens." There was little opportunity for the evaluator to question the appropriateness of the objectives themselves.

As can be seen from this overview of evaluation models, the field of educational evaluation during this period represented very general aids to conceptualizing evaluation. These aids both guided and in some ways limited what could be evaluated. Other more technical models embodying more specificity have been developed for highly specialized, idiosyncratic applications but these have had limited generalizability across educational settings. (See, for example, Borich, 1974, for a review of some of these more specific models.)

Table 1.2 summarizes these three types of educational evaluation models. It can be noted that the purposes of the evaluation and the implied role of the evaluator differ little across them. However, as one moves from general purposes to specific evaluation activities, the diversity in these procedures becomes greater. The point to be emphasized from consideration of Table 1.2 is that the role of evaluator changed little during the early development of the field. During this period these models represented educational evaluation's best efforts to meet the needs of program decision makers. We will have more to say about more recent efforts in the field of educational evaluation in subsequent chapters.

SOME GENERAL COMMENTS ON MODELS

Stake (1981) has advocated use of the term *persuasion* instead of *model* when referring to evaluation approaches. While we concur with this distinction, we would also argue for a better understanding of the concept of a model in evaluation. The desire among evaluators to develop and use models (e.g., the CIPP Model, the Discrepancy Model, the Stake Model) is understandable since there is the hope that once a model is established it can be used in a wide variety of contexts. However, as Stake (1975b, 1981) has suggested, evaluation does not always work that way. The techniques

Table 1.2
Summary of Three Categories of Decision-Oriented Models of Program Evaluation

	Educational psychology models	Educational decision models	Educational science models
General purpose of evaluation	To determine discrepancies between objectives and outcomes and between the intended and actual program implemented	To make better, more defensible decisions	To determine causal relationships between inputs, program activities, and outcomes
Implied role of evaluator	Information provider, informs decision makers of program discrepancies, may make recommendations based on judgments	Information provider, serves the decision maker	Technical expert providing statistically manipulated data (usually of the significant–nonsignificant variety) to decision makers
Importance of objectives	Very high; objectives are the touchstone for evaluation	High; decision maker informs evaluator of objectives, who in turn gathers data and reports back to decision maker	Low; purpose of evaluation is to determine causal links; deals with variability in outcomes rather than number of scores above some cut-off, etc.
Types of proposed evaluation activities	Program description, determining logical and empirical contingencies, noting discrepancies, possible cost–benefit comparison to alternative program	Evaluation of context, input, process and product. Heavy emphasis on data gathering and integrating divergent sets of data	Experimental design, setting up study to meet statistical requirements. Data collection and analysis
Methodological approaches emphasized	Observation of program, psychometrics, criterion and norm referenced testing, parametric statistics (usually group comparisons)	Both quantitative and qualitative approaches; dictated by needs of decision makers	Quantitative methods; usually parametric procedures utilizing correlations and tests of significance

and methods brought to bear in an evaluation are a function of the problem, the clients for whom the evaluation is being conducted, the values inherent in the program context, and the amount of time and money that can be devoted to it. While evaluation is not an art form, evaluation models can communicate only a relatively small set of categories, stages, and constructs that might be useful in planning an evaluation. The problem of choosing an evaluation model derives, in part, from a somewhat natural tendency to see an evaluation model as more than it is. As Stake (1981) points out, evaluation models have been mistaken as methodologies for actually conducting evaluation, instead of as persuasions or frameworks within which more specific constructs and methods must be placed. Although evaluation models do not provide answers, they do provide guidelines or heuristics for thinking about how an evaluation could be conducted. This heuristic role for models, which has not always been appreciated in evaluation theory and practice, has been described by Kac (1969):

> The main role of models is not so much to explain and to predict—though ultimately these are the main functions of science—as to polarize thinking and to pose sharp questions. Above all, they are fun to invent and to play with, and they have a peculiar life of their own. The "survival of the fittest" applies to models even more than it does to living creatures. They should not, however, be allowed to multiply indiscriminately without real necessity or real purpose [p. 699].

Hesse (1976) divides models into models *in* science and *of* science. According to Hesse models *in* science (e.g., model of the atom, model of crystal structure, model of conflict) are inductionistic. They yield observations of naturally occurring phenomena that unambiguously support or refute the nature of the model. Models in science are *precise*. That is, measurements are devised to describe the phenomena of interest and the model is basically quantitative in nature. Also, models in science are *specific*. They deal with only a select number of phenomena, purposefully avoiding summary or aggregate characteristics of an event or entity. Furthermore, models in science are *verifiable*. Hypotheses can be posed to check the precision of the model and the empirical evidence accumulated, which eventually determines its accuracy and usefulness.

Models *of* science (e.g., positivist model, hypothetico-deductive model, evaluation models) are deductionistic; they are theoretical or philosophical persuasions within which observations are interpreted. For these models, there is nothing outside of the model itself from which we can unambiguously determine what exists or what is true. Models of science are not precise, specific, or verifiable. They usually do not contain the means with which the phenomena could be measured. They often deal with a wide variety of phenomena by summarizing or aggregating variations into broad clusters or categories of events. It is difficult to check the accuracy or usefulness of the model by posing hypotheses and gathering empirical evidence.

This latter classification fits well with the way the concept of a model should be used in the field of evaluation. We even suspect that it represents the sentiments of the model builders themselves. Unfortunately, we have often mistakenly utilized evaluation models as models *in* and not *of* science.

How then should evaluation models be used? A limitation of models is that they impose implicit restrictions on the types of rival hypotheses or alternative explanations that can be expressed. Models divide the world of programs into predetermined categories (e.g., context, input, process, product) that can readily be taken to mean that everything needed to understand program effects is included within the domain of the model (and the categories, stages, or steps it contains). Properly used, an evaluation "model" is a heuristic for organizing thinking about an evaluation, not an algorithm or paradigm for actually conducting an evaluation.

Models may restrict not only the questions that can be asked, but the viewpoints from which questions may emerge. Models often represent only a single viewpoint. They conform to the viewpoint of administrators, managers, and supervisory personnel who make "on line" decisions about programs, or the viewpoint of the key constituents who have a stake in a program, or of two opposing sides arguing about the quality of a program. We might ask if any single model can address the full range of questions that is likely to be asked about a program by various constituents. Multiple questions from diverse viewpoints usually cannot be raised within the context of any single model.

Some concepts broader than models, such as philosophies or theories of what should be looked for in a given context, might best guide our selection of specific evaluation techniques and procedures. Some of Stake's persuasions (1981), House's taxonomy of evaluation types (1980), or the more generally conceptualized perspectives of decision-oriented, value-oriented and systems-oriented evaluation, to be presented later in this volume, are representative of these guideposts.

The Growth of Decision-Oriented Program Evaluation in the Field of Human Services

While the historical roots of program evaluation run deepest in the field of education, the current clamor and concern over evaluation often stems from other human service areas such as corrections, family, financial, health, and mental health interventions. It is in the social service domain of human services that the field of evaluation is achieving its most advanced stages of sophistication while simultaneously becoming less and less able to cope with the demands being placed upon it. This is not to imply that the field of education has been static and evaluation problems there pale in comparison with those in human services. Rather, the tremendous increase in emphasis, growth, and complexity of our human service programs has outstripped our ability to provide meaningful evaluative data in these areas. The field of education has not been nearly so tempestuous. A brief historical sketch of the development of human service will help illustrate the current state of affairs.

HISTORY OF HUMAN SERVICE PROGRAMS

Until the 1930s human service programs were poorly organized. The services provided were aimed at special populations and were usually of a limited nature. These programs tended to be local and funded as charities. With the exceptions of the Emancipation Proclamation in 1863, and the child welfare legislation of 1910 and 1912, the federal government steered clear of social policy. "Rugged individualism" and "paying your own way" were predominant social values of the day.

The Depression of the 1930s portended major changes in the federal government's role in social policy. In light of the massive failure of the country's economic structure, the federal government assumed major responsibility for the welfare of the general public. The most notable thrusts of this movement were the creation of a large number of federal jobs through the Work Projects Agency (WPA) and the Civilian Conservation Corps (CCC), the development of the Office of Education, and the establishment of a pension system for the elderly through the Social Security Board. Even more important than the programs sponsored during this period was the emerging trend of federal intervention in public welfare.

Another major influence on the development of human services was the geographical and social mobility that occurred just before and after World War II. This trend, established in the early 1930s as an attempt to maintain societal stablility, became the justification for further involvement in social welfare by the federal government as industrialization signaled the end of the extended family and its ability to deal with the social problems of its members. This movement culminated in the formation in 1953 of a cabinet level department called the Department of Health, Education and Welfare.

The civil rights movement of the late 1950s further increased the federal government's involvement in social welfare not only through specific programs, but also in response to concerns that responsibility for implementation of new social policy could not be left in the hands of local authorities. The civil rights movement stands as an impressive forerunner of the self-examination and questioning of values that characterized American society throughout the 1960s.

Hersh (1972) put forth the notion that this period was characterized by an increased concern for the disadvantaged, oppressed, and powerless individuals and groups in our society. Certainly, the community mental health legislation of 1963, the civil rights legislation under both the Kennedy and Johnson administrations, the medicare legislation of the mid-1960s and the previously mentioned Elementary and Secondary Education Act of 1965 reflected the federal government's readiness to respond to these social concerns.

The reader is spared an attempt to detail the mushrooming of human services since the mid-1960s. Summarizing Attkisson and Broskowski's (1978) figures on the growth of HEW, it is evident that the agency's overall budget has increased almost 16-fold between 1954 and 1975. Federal per capita expenditures for social welfare rose from $33 per person in 1954 to $400 per person in 1972. HEW personnel increased from 25,000 in 1953 to 78,000 by 1970. By 1975 this figure had increased to 123,000. By 1975 HEW constituted the largest single agency in the Federal government, accounting for one-third of the federal budget. See also Attkisson and Broskowski (1978) for a review of the growth of human service programs.

THE CATEGORICAL PROGRAM APPROACH

A major difficulty with the growth of federal service programs has been the manner in which new programs are justified and brought to fruition. Traditionally, rather specific programs have been developed to meet narrow needs. This practice has been labeled the categorical approach to human service (cf. Attkisson and Broskowski, 1978; Demone, Schulberg, and Broskowski, 1978; Spencer, 1974; Weiss, 1972c) and has resulted in large numbers of related but unconnected educational, health, and social welfare service agencies existing in any given community in the country. These specialized programs usually have a narrow definition of services and a limited clientele, and are funded by some combination of federal, state, local, and private sources.

As with other issues we have discussed in this chapter, categorical social programming was not a planned strategy. Rather, the growth of categorical programs is seen as an ill-planned phenomenon, guided more by immediate convenience, practicality, and political reality than by careful planning and foresight. Categorical program sponsors in the Congress could reap some political "hay" by addressing particularly pressing social problems. Categorical program funding allowed administrators of these programs more discretion over funds. Related to this was the fact that some administrators were concerned that their service specialty might not be allowed to fully address societal needs within the aegis of some more general or related agency. Special interest groups favor a categorical approach because of the increased possibility of their specific needs being addressed. Perhaps more importantly, it is easier to react to the demands of special interest groups than it is to determine overall priorities before allocating resources. However, such strategies will almost always yield remedial rather than preventative programs (Demone et al., 1978).

Problems arising from this categorical approach to service delivery are overwhelming. Attkisson and Broskowski (1978) point out that competition for funds and authority has impeded an integrated approach to service delivery. They conclude that the more than 300 programs of HEW are "frequently judged to be fragmented, duplicatory, inefficient, at times contradictory and generally unmanageable [p. 15]." The duplication of management overhead costs of different agencies providing the same or similar services to an overlapping clientele has left the human service system extremely complex, rigid, inefficient, ineffective, and productive of reams of virtually unintelligible data and red tape.

In addition to these problems, Spencer (1974) documents the plight of a prospective client who tries to deal with this fragmented collage of human services. He reports that 60% of the qualified clients seeking social services never receive them. Furthermore, the probability of the effective referral of a client from one agency to another is only 17%. He concludes that since 86% of those seeking services are in need of more than one service, the probability of getting the first service is 40%, and the chance of a successful referral for other needs are so slim that the individual in need of several services has very little chance of receiving them. Spencer's (1974) data support the contention that those most in need of services are those least likely to get them. The following quote from Attkisson and Broskowski (1978) summarizes well the problems the categorical approach poses for evaluation: "We have reached the point where legislators, administrators, providers, and clients are incapable of determining what is being done, by whom, for whom, and at what cost [p. 15]."

NEW FEDERALISM AND THE HUMAN SERVICES CONCEPT

Due in part to the early-appearing problems with the categorical approach to programs, and in part to the stern conservatism and fiscal policies of the Nixon administration, new directions were being sought in the early 1970s in reaction to the tremendous proliferation of programs under the *Great Society* administration of Lyndon Johnson. Dubbed *New Federalism*, this policy produced several changes, with the goal being to shift Federal responsibility for human welfare from a service strategy to an income strategy (Nathan, 1974). This income strategy generally involved providing cash and in-kind benefits (e.g., food stamps, medicaid) to the needy rather than specific services. A second major thrust of New Federalism was revenue sharing which provided an avenue for shifting control over discretionary social program funding from the federal to the local level.

While it is too early to determine with any certainty the future political climate on this issue, it can be expected that the growth of new programs will slow, with a shift in emphasis toward integration of already existing services and away from categorical program funding. Models for program integration are rare, owing in part to a failure on the part of human service administrators to think in terms of system concepts. Some inroads to integration of services have been made. The short history of this movement is reviewed by Attkisson and Broskowski (1978). Spencer (1974) provides a useful framework for system integration and proposes the concept of an *integrator* by which such a confluence of programs may be attained. He suggests some new tools (program diagramming or modeling for example) to aid in achieving this integration and discusses a prototypical integrated program in a metropolitan area.

THE CURRENT ROLE OF EVALUATION IN HUMAN SERVICES

Human service program evaluation began primarily as an extension of educational evaluation, borrowing freely the concepts, methodology, and even personnel of the older enterprise. This trend is particularly evident in mental health evaluation. The concept of behavioral objectives was parlayed into the notion of client goals. Other basic similarities between the two fields suggesting common roots include similar process evaluation strategies, reliance on standardized test scores as outcome measures, and the use of experimental and quasi-experimental designs to assess program impact. It should be mentioned that many of the problems and criticisms cited under educational evaluation also apply to human service evaluation. It should also be mentioned that human service program evaluation has contributed new concepts and techniques to the general field of program evaluation. Cost-effectiveness and cost-benefit analyses have been successfully adapted from the fields of economics and operations research and applied in mental health and other human service settings (Carter and Newman, 1976; Hagedorn et al., 1976; Sorensen and Grove, 1978). Human service evaluation has also taken greater initiative in the areas of epidemiology and management information systems (Flaherty and Morell, 1978).

Furthermore, the field has developed a broad array of outcome measures more suited to its needs and constraints than those conceptually similar outcome measures borrowed from education. These include the Global Assessment scale (GAS), the Symptom Checklist (SCL-90), the Automated Tri-Informant Goal Oriented Progress Note (ATGON), the Client Satisfaction Questionnaire, and the Katz Adjustment Scale (KAS), to name only a few. The field has also developed an array of heuristics to aid in administering program evaluation, utilizing management information systems, determining patterns of use, clinical audits, and self-evaluation. (See Lund, 1978, for a humorous but informative view of these recent developments and Hagedorn et al., 1976, for a comprehensive survey of this developing methodology.)

Despite this developing methodology, evaluation in human service contexts remains decidedly decision-oriented. Whether part-and-parcel of the legacy from educational evaluation or (more likely) the result of its own pressures of a practical, utilitarian, convenient, or political nature, human service program evaluation appears to be guided by the same decision-oriented philosophy as is found in education. There can be little doubt that the same dangers and caveats apply.

Nor can there be much doubt that the decision-oriented perspective predominates in human service. Consider the following quote by Lund (1978):

> I have referred to program evaluation as a "service." I mean that quite literally. The evaluator is and should be in a service role providing needed information to clinicians, managers and other program constituencies. If he fails to actively provide accurate, timely and interpretable information to management, he fails in the achievement of his own core mission as an evaluator [p. 39].

While not all writers define the province of human service evaluation as narrowly as does Lund, virtually all acknowledge the decision-oriented perspective that guides their activities (cf. Anderson and Ball, 1978; Rossi, Freeman, and Wright, 1979).

Given current accountability pressures in human service, many in the field are implicitly or explicitly calling for evaluators and the field of evaluation in general to play the integrator role suggested by Spencer (1974). Attkisson, Brown, and Hargreaves (1978) propose a three-dimensional model of evaluation capability that casts the functional role of the evaluator into the continuum depicted in Table 1.3.

Attkisson, Brown, and Hargreaves point out that often evaluators are not allowed to participate in or even have access to decision making and planning. In the statistician role the evaluator is no more than a sophisticated clerk who manages, summarizes, and reduces large amounts of data. The evaluative researcher role does examine relationships among different variables, particularly cause–effect relationships. However, the time lag between data gathering and report writing is usually so long that the report has little value for ongoing decision making. While the evaluative technician role

Table 1.3
Evaluator's Functional Role Continuum

Evaluator as statistician	Evaluator as researcher	Evaluator as technician	Evaluator as integrator, coordinator, and decision maker
Compiles facts from natural data sources for funder-report purposes	Attacks problems of applied research interest	Reports potentially valuable information to management but does not participate in decision making	Actually becomes involved in decision making; ensures evaluative data input into the decision making process

does allow access to the the decision-making process, continued input from the evaluator is not usually sought. Thus, the evaluator's knowledge and understanding of the organization and the relationship between evaluative data and policy are ignored. The integrator role helps ensure that evaluative information receives "direct, reasonable and participative advocacy in relation to other influential factors such as community politics, service provider allegations, administrative predilections, and financial constraints [pp. 82–83]." These writers propose three potential linkage methods for integrating evaluation and administration.

Similarly, Cronbach et al. (1980) point out that "The theory of evaluation . . . has been developed almost wholly around the image of command [p. 84]." This is the view that there is a clearly identifiable decision maker on the horizon who makes decisions on the basis of informative input, including that provided by the evaluator–statistician. These writers point out that the actual context in which evaluation occurs is not consistent with the *context of command* assumed to exist in virtually all program settings. It is our contention that the assumption of a context of command (Cronbach's term) or that evaluation has been decision-oriented (our term) has not been a result of accident or serendipity. Rather the evolution of this state of affairs has resulted from the belief that evaluation's central role is to respond to the pragmatic demands of the marketplace and to the needs of those in most immediate control of the program. Hence, more complex decision-making patterns (Linblom, 1965, p. 29ff) and more holistic criteria of program effectiveness were obscured by the image of a more definitive but overly simplistic command structure.

The work by Attkisson et al. (1978) and Cronbach et al. (1980) points out that evaluative data compete with other sources of information in the decision-making process. Some of the competing sources are quite strong-armed (for example, extant political presses) and simply overpower the information generated by a more clerical, decision-oriented perspective on evaluation. It seems incumbent upon the field of human service evaluation to devote itself to the integration of evaluation and administration, and to the development of new skills, techniques, and strategies that yield more relevant information than has often been offered by evaluation in the past; and to develop more appreciation of the context in which programs (and thus evaluation itself) operate.

We have emphasized the need for more robust evaluation strategies because of the political and resource constraints requiring more responsive, efficient, and effective human service programs. The shift from centralized to local control of the budget and a tighter budget has created the need for better evaluative information, better accountability of resources available and utilized, and input into integrative planning. Attkisson *et al.* point out that these developments have clouded the issues of funding and lines of authority. Similarly, the importance of the policy shaping community and the need for evaluation to address the needs of this group rather than some mythical lone decision maker suggest basic reform in the way evaluation has traditionally viewed itself (Cronbach *et al.*, 1980). Given the state of flux in which human service program evaluation finds itself, one questions the ability of the field to play the integrator–educator role called for in human service programs. Attkisson *et al.* expressed a similar view. "The principle issue we must address, however, is whether the evaluative field is currently mature enough to respond to the hefty expectations being placed upon it by law, by administrators, and by evaluators [p. 60]." They go on to cite the following limitations of the field of evaluation, which impede rapid expansion of present evaluative capacity:

1. Excessive expectations for evaluative information
2. Limited funding in human service and specifically in evaluation
3. A lack of adequately trained evaluators
4. Poor information systems and computer support facilities available to local agencies
5. Differences of opinion about the appropriate roles of evaluators.

At the same time Flaherty and Morell (1978) cite four reasons for the *growth* of evaluative capacity.

1. Accountability requirements have toughened, stimulating the development of better methodology
2. Social scientists have become more concerned with social behavior.
3. Scarcity of resources in the traditional social sciences has brought these thinkers into more applied settings
4. The potential for expansion of social science methodology has acted

as impetus for the development of evaluation methodologies better suited to the demands of human service delivery.

So where does that leave program evaluation in the fields of human service? The evidence cited here points toward either the hot seat (evaluator as integrator) or the basement (evaluator as clerk). The need to break from the traditional decision-oriented perspective in favor of a more integrated systems view of itself and the context in which it operates characterizes the current status of human service program evaluation. Thus, evaluation needs to: (a) increase its effectiveness by providing more relevant and timely data, and (b) integrate itself more closely with the political processes through which decisions emerge. Although solutions to these problems are interdependent, reform will not be easy.

Summary

There is little doubt that decision-oriented evaluation has come to dominate the field of program evaluation in both education and human service. Given the practical nature of decision-oriented models this predominance is easily understood. While decision-oriented evaluators purport to do the right thing, this conceptualization of evaluation defers power, authority, responsibility, and even technical expertise to the decision maker. This allows the decision maker to dictate what will be evaluated, how the evaluation will be conducted, and how results will be interpreted. It is this approach to evaluation that we call the decision-oriented perspective. This perspective is one in which decisions about what will be evaluated, how the evaluation will be conducted, and how the results will be interpreted are made with little or no input from the evaluator. We contend that such contexts reduce and can even subvert the role of evaluation. It is only by recognizing and by treating limits on the freedom of the evaluator as "data" about the program that one comes closer to relating program results to values, an activity implicit in the word *evaluation*. Although the intents of prominent evaluators who originally pioneered decision-oriented evaluation were to foster the utility and validity of the evaluative enterprise, their legacy has also served to limit the professional role of the evaluator and may have unwittingly compromised the evaluators' responsibility to the profession and the public to be served.

2

The Decision-Oriented Nature of Program Evaluation

We have reviewed the development of program evaluation in the field of education and the general field of human services. We can conclude from these reviews that evaluation has developed and been guided more by practicality and convenience than by anything else. This unplanned growth has resulted in an eclectic bag of evaluation models and techniques that has produced few philosophical or theoretical underpinnings for the field as a whole. Thus, guidelines for selecting among the vast array of techniques available and clarity as to appropriate roles for evaluators have not evolved. Instead, evaluation has evolved rather narrowly as a technical aid to decision making, as one source of informative input to the evaluative process. Defining evaluation this way has kept the emerging field out of the political wringer and has given evaluation an unassuming, nonthreatening niche within education and human service, thus ensuring everyone's continued pleasant disposition and employment.

We are not saying that evaluators chose to be decision-oriented for these reasons. On the contrary, we are saying that by applying what they knew (usually social and behavioral science statistics) to the questions being posed by decision makers without considering the larger context of their work or the development of their profession, evaluators have unintentionally created a philosophy by default. This philosophy has predisposed the evaluator to the compromising role of providing information to those higher up the hierarchial decision-making ladder. Some resent this role for evaluators, believing they can be, should be, and in many cases are capable of contributing more to program planning, management, and evaluation.

The Political Context of
the Decision-Oriented Approach

By unwittingly abdicating judgment (and responsibility) to others, evaluators have created the potential for serious abuse of the evaluative enterprise. Political press, more than anything else, has contributed to this state of affairs (Brickell, 1976; Cronbach et al., 1980; Horst, Nay, Scanlon, and Wholey, 1974; Suchman, 1972; Weiss, 1972c).

Suchman (1972) cites the following as possible misuses of evaluation:

1. Eye-wash—an attempt to justify a weak or bad program by deliberately selecting for evaluation only those aspects that "look good" on the surface. Appearance replaces reality.
2. White wash—an attempt to cover up program failure or errors by avoiding any objective appraisal. Vindication replaces verification.
3. Submarine—an attempt to "torpedo" or destroy a program regardless of its effectiveness. Politics replaces science.
4. Posture—an attempt to use evaluation as a "gesture" of objectivity or professionalism. Ritual replaces research.
5. Postponement—an attempt to delay needed action by pretending to seek the "facts." Research replaces service [p. 81].

Brickell (1976) provides 10 real life examples of these evaluative misuses that are both humorous and informative. They illustrate the fact that within a decision-oriented perspective, the reality of the situation is that the evaluation funder or program manager can be expected to use whatever power he or she may have to influence evaluation data, results, and conclusions. The following quote from Brickell (1976) illustrates this point.

Are there external political factors that influence the role and methodology of evaluation? You bet there are. And they are powerful.

Because the seeking, the winning, and the exercising of power are a prominent part of life within an institution as well as between institutions, internal evaluation staffs are just as subject to political influences as external evaluation staffs. Perhaps you have noticed that. Sometimes political forces control the populations we can sample. Sometimes they limit the data we gather. Sometimes they shape our instruments. Sometimes they influence the designs we can use. Sometimes they guide our interpretations. Sometimes they shape our recommendations. Sometimes they touch the wording of our reports. And they always influence the impact of what we recommend.

I think I have never written an evaluation report without being conscious of the fact that what I say will be used in the winning and the exercising of power, that my findings are going to be lined up on one side or the other of a contest that somebody else has already set up, and that jobs are on the line—maybe my own job [p. 5].

One of the most definitive accounts of the political context in which evaluation occurs has been that recently offered by Cronbach and his colleagues in the Stanford Evaluation Consortium (Cronbach et al., 1980). Integrating the works of a number of notable writers in the field of evaluation, they suggest that evaluation be reformed so as to be more in tune with its political context. As we discussed in Chapter 1, these writers present considerable evidence and convincing dialogue supporting the notion that the "context of command," traditionally assumed to characterize decision making in programs, may in fact be a rather tenuous assumption. By context of command it is meant that there is an identifiable individual on the horizon who is in complete control but who needs information with which to act. Once provided with information, this decision maker issues orders and directives to carry out the decision made. It is further assumed that orders and directives are carried out precisely and that this manager or decision maker is responsible for the decision made. Cronbach et al. point out that the view of a lone decision maker is not congruent with the way most programs evolve and proceed. Rather, they argue that most social action evolves in a pluralistic community referred to as the *policy shaping community* (PSC). Cronbach and his colleagues describe the evolution of decisions in the PSC this way:

> Most action is determined by a pluralistic community, not by a lone decision maker. Parties having divergent perceptions and aims bear down on lawmakers and administrators. Information can change perceptions but is unlikely to bring all parties to agree on which facts are relevant or even on what the facts are. If the term *decision* is understood to mean formal choice at a particular time between discrete alternatives, decision making is rarely to be observed. When there are multiple participants with multiple views, it is meaningless to speak of one action as the rational or correct action. The active parties jockey toward a politically acceptable accommodation [Cronbach et al., 1980, p. 84].

These writers describe the PSC as a large group of people consisting of everyone who may have an interest in the program. Decisions are seen to evolve; there is more a "drift toward decision" by this community than there is an identifiable, clear-cut decision for action. This drift toward decision is the result of the processes of negotiation and accommodation among members of the PSC. Because of the inherently political nature of such a process, the "decision" that eventually emerges is apt to be a piecemeal adaptation to a problem rather than a broad change in any particular system or systems. A compromise struck among divergent interests seldom portends radical or far-reaching change. Cronbach and his colleagues draw effectively on the work of Lindblom (Braybrooke and Lindblom, 1963; Lindblom, 1965, 1972; Lindblom and Cohen, 1979) and his conceptualization of disjointed incrementalism in making the case that evaluation occurs within a context of accommodation, not a context of command as has been traditionally assumed.

The reform suggested by Cronbach and his colleagues emphasizes evaluation's role as a multipartisan educative function. Consider the following, which are a few of these writers' theses bearing on this point.

- Major evaluations should have multiple sponsorship by agencies with different perspectives.
- Those who shape policy should reach decisions with their eyes open; it is the evaluator's task to illuminate the situation, not to dictate the decision.
- The evaluator is an educator; his success is to be judged by what others learn.
- When programs have multiple and perhaps dissimilar outcomes, comparison is invariably judgmental. No technology for comparing benefits will silence partisan discord.

Of course, the precise effects of erroneously assuming a context of command by the field of evaluation cannot be specified; but it is tempting to speculate about these effects in terms of utilization problems, timeliness, report style, impact, and similar issues whose relevance to political considerations are obvious. Perhaps less obvious are the restrictions placed on evaluators which are part-and-parcel to assuming that evaluation is decision-oriented, or in Cronbach's terms, assuming a context of command. We will address the methodological implications of these often-made assumptions subsequently in this chapter. But there are some additional political considerations bearing on evaluation that should be noted.

WHOSE FAULT IS IT?

Horst, Nay, Scanlon, and Wholey (1974) question whether the organizational environment in which evaluation finds itself will actually permit meaningful evaluation to occur. They point out that attempts to redress the major causes of evaluation problems have had little impact on program performance and that most social programs are administered by "pseudomanagers." Pseudomanagement refers to the occurrence of management activity independent of actual program results. Pseudomanagement is judged effective if it pleases the continually changing administrators and politicians to whom it is responsible.

Horst *et al.* propose a preassessment of a program's evaluability before actually beginning the evaluation. They suggest three dimensions of evaluability: (*a*) measurability of program goals, objectives, and activities; (*b*) faulty logic in program assumptions; and (c) existence of a management vacuum. They propose that by following their guidelines a program may be judged "evaluable," "potentially evaluable with further program or management definition," or "not evaluable." They suggest that this rating of a program should be communicated to policymakers and program managers close to the program *and* to the upper echelons of agency management so everyone will know what programs are or are not evaluable and why. Acknowledging the risk of these activities, they suggest that such a preassessment could preclude poor evaluation and recriminations against (and disrespect for) evaluators, as well as contribute to ultimate program improvement.

LET'S RUN IT LIKE A BUSINESS!

Several writers in the fields of evaluation and public policy have suggested that we run our human service programs much as we would a private, profit-making enterprise. Linowes (1974) notes some interesting differences between private, profit-making agencies and social service agencies. He points out that if a company sinks substantial resources into an ill-fated enterprise, the company suffers a major loss. On the other hand, social service agencies are not forced out of business if not doing well. In many instances "doing well" is not even defined. Corporate executives are usually much more accountable for their performance than are government bureaucrats.

Linowes suggests that the superior success of business in meeting objectives (relative to the performance of social agencies) can be attributed to the application of a total concentrated management approach. He suggests the use of such an approach to the management of social programs, utilizing, where applicable, highly developed and sophisticated management strategies that have guided growth and profitability in business and industry.

Linowes proposes Socio-Economic Management (SEM) as a means of identifying exemplary programs and doing away with unworkable, unmanageable, and unaffordable programs. He delineates 10 rules of SEM:

1. Tie standards and goals to proven human needs.
2. Apply funding by results.
3. Use discretionary funding as incentives.
4. Use multidisciplinary planning.
5. Set up social profitability audits.
6. Establish public visibility.
7. Prune and restructure for dynamic growth.
8. Vary the input mix.

9. Stir up competition.
10. Fix responsibility for implementing SEM.

A conceptually similar approach is suggested by Broskowski and Driscoll (1978). They took from the futurist corporate designs forecast by the noted corporate planner and educator Jay W. Forrester those attributes predicted to be characteristic of profit-making enterprises in the future and suggest their use as guidelines or goals for the organizational development of human services. We are not sure, however, that the application of business principles is as simple and straightforward as suggested by either Broskowski and Driscoll or by Linowes. Cronbach and his colleagues (1980) have pointed out that clear cut decision making (effective command) is not as prevalent in government-sponsored services as it is in private industry. If Cronbach's view of accommodation and negotiation is a more accurate portrayal of the procedural context within which programs evolve and operate, then models derived from organizational contexts of command are not likely to be applicable. Thus, run-it-like-a-business arguments often deny the political context in which government-sponsored programs operate or are naive about these considerations all together.

EVALUATION IN THE BROADER CONTEXT

The issues of evaluability–pseudomanagement raised by Horst et al. (1974) and the social service-as-business arguments just reviewed are interrelated. Some concepts and notions borrowed from Thompson (1967) will help illuminate this and other conceptual relationships in the arena where evaluation occurs. Thompson proposed two continua or dimensions that dictate the kinds of evaluation that go on in organizations. One dimension is the completeness of knowledge about cause-and-effect relationships. We could call this dimension *information grasp.* This is synonomous with the Horst et al. dimension of the strength or faultiness of the logic of assumptions underlying program structure. Thompson's other dimension is the degree of crystallization of standards of desirability. We could call this dimension *knowing what we want.* These dimensions are cast on a 2 × 2 matrix in Figure 2.1.

According to Thompson, when information grasp (i.e., empirical knowledge of cause-and-effect) is high, and when one knows what is valued (i.e., standards of desirability are well-crystallized), an organization will assume a maximizing rationale for evaluation whereby maximal attainment of objectives per unit cost (i.e., efficiency) is emphasized. When knowledge of cause-and-effect is low but standards of desirability are well-crystallized, a *sufficing* approach (Simon, 1969) to evaluation is promoted, whereby acceptable or desirable states are sought and criteria play an instrumental role

Standards of Desirability

		Highly Crystallized	Not Crystallized
Knowledge of Cause- and-Effect	High	Maximizing approach; emphasis on efficiency criteria. Characteristic of profit-making agencies. Use of cost–benefit methodology. I	Evaluative criteria derived from wants and needs of special interest groups (usually target population of program or other special interest group). II
	Low	Sufficing approach; emphasis on successive approximations of the desired state of affairs. Characteristic of social service agencies. III	Evaluative criteria derive primarily from political, administrative, and personal presses. IV

Figure 2.1.   A 2 × 2 matrix of two dimensions relevant to the conduct of evaluation.

in evolving approximations of what is most desirable. Emphasis is on improvement rather than maximization. When standards of desirability are not well-crystallized, the values and standards of particular interest groups, politicians, administrators, or the public become the criteria by which the program is evaluated. Within this category, we might speculate that when knowledge of cause-and-effect is high, special interest group values predominate. When knowledge of cause-and-effect is low it would appear that political feasibility predominates.

Several aspects of Figure 2.1 warrant comment. The lack of standards of desirability in Quadrants II and IV creates monumental problems for evaluation. With poorly crystallized standards of desirability, judgments of worth are not possible; evaluation can only be descriptive. Furthermore, decisions are not possible without criteria. When standards are unclear and criteria for judging success ever-changing, management simply cannot occur. When management does exist in this situation (and it invariably does; no one wants to admit to chaos!) it is usually for the sake of appearance. This concept is similar to the Horst *et al.* (1974) description of the pseudomanagement existing in federal bureaucracies. Pseudomanagement may be the best that can be hoped for when standards of desirability are not well-crystallized.

On the other hand, the application of a maximizing approach to evaluation, implicit in the suggestions of Linowes (1974), Broskowski and Driscoll (1978), and others, would require a high level of knowledge about cause-and-effect *and* highly crystallized standards of desirability (Quadrant I in Figure 2.1). The application of business principles to social service programs may not be as simple and straightforward a process as these writers imply. Major empirical knowledge of cause-and-effect relationships in human services and methodological tools to obtain crystallized standards of desirability are now lacking.

The question becomes one of determining where in the matrix any particular program falls. Suchman (1972) implies that evaluation in the public sector falls into Quadrant III. However given the problems encountered by evaluation, one wonders if most federally funded programs do not actually fall into Quadrants II and IV. The following relationship is hypothesized to exist between the matrix in Figure 2.1 and the Horst *et al.* classifications of evaluability. Evaluable programs, in Horst *et al.*'s terms, exist primarily in Quadrant I and, to a limited extent, in Quadrant III. "Potentially evaluable" programs are found almost exclusively in Quadrants II and III. "Not evaluable" programs will be found primarily in Quadrant IV, although some may be found in Quadrant II.

The purpose of this discussion has been to show how critical factors influencing the context of evaluation interact to constrain evaluative activity. Although the conceptualization of the evaluative context portrayed in Figure 2.1 may be overly simplistic (major dimensions represent a continuum, not just extreme end points), it does illuminate the issues of program evaluability and addresses the flexibility of human service programs to adopt hard-nosed, efficiency-oriented management strategies from management science. The issues are neither simple nor clear cut. However, by taking a step back and considering the larger context, one can gain some insight into such problems.

Methodological Flaws of the Decision-Oriented Approach

The political and contextual problems just mentioned both give rise to and are a result of the decision-oriented approach to evaluation. However, this is only half the story. On the other side of the coin are substantive methodological issues that, although generally acknowledged by other writers in the field, have not been explored, examined critically, or addressed adequately in the literature (with a few notable exceptions). These issues involve the applicability of social science research methodology to evaluation settings.

The reader might at this point take exception to our contention that the issue of the applicability of experimental design to evaluative inquiry needs further debate. Certainly, one can find ample verbiage devoted to the topic of evaluation versus research in articles (cf. Guba, 1969), books (cf. Worthen and Sanders, 1973), and professional newsletters (cf. Sommer, 1977). Far from beating a dead horse, we believe closer examination of this nexus of considerations leads to a more complete understanding of the limitations of decision-oriented methodology. This issue goes beyond arguments of evaluation versus research. We contend that the field has in general paid lip service to this difference, acknowledging some limitations of a research approach to evaluation while doing very little to change its actual behavior.

In light of the current state of affairs in evaluation practice, with the seeming pressures to become more rigorous and facilitative of information exchange and intelligent decision making, this status quo in behavior is difficult to understand. One need not look far to find the undergirding role played by the scientific method and the statistical methodology of the social sciences. Certainly they predominate in technical and published evaluation reports. The training of evaluators continues to be heavily, if not exclusively, steeped in traditional statistical designs as a way of measuring project impact (cf. Horst, Tallmadge, and Wood, 1975).

Another index of our nearly exclusive reliance on the scientific method and its attendant methodology is the difficulty we have in even acknowledging the differences between evaluation and research. We occasionally take time to note the points mentioned by Stufflebeam et al., 1971 (see pp. 12-13, this volume), but then go about our evaluative business as if we were researchers, totally unaware of our own caveats. We still use the terms *applied research, applied social research,* and *evaluation research* when describing our activities. A cursory review of books on evaluation almost invariably reveals some research-related moniker in either the title or subtitle of the volume. Perhaps more deceptive is the practice, among research-oriented evaluators, of putting forth decidedly decision-oriented views without revealing their bias toward research. The recent book by Rossi, Freeman, and Wright (1979) is but one example of the emphasis placed on randomized, controlled experiments as the optimal means of assessing the effects of an intervention. There can be little doubt that the scientific method with all its trappings has become the modus operandi for evaluators.

Nor can there be much doubt that this reliance on social science methodology is intimately tied to the evolution of decision-oriented evaluation as we have presented it. We discussed earlier the legitimacy borrowed from the harder sciences through using its technology, the sanctuary provided harried evaluators by go-no-go statistical decision rules within defined confidence intervals, and the fact that much evaluator training has been, at best, in the field of social science statistics. The use of such methodology certainly fits the requirements of convenience and practicality, the primogenitors of the decision-oriented approach.

Even the very nature of social science statistics fits most comfortably with a decision-oriented perspective. The whole point of parametric statistics is to be able to make decisions about some larger population based on statistics obtained from a sample drawn from that population. Of course, data *reduction* is one of the most seductive aspects of statistics. Disregarding what may be lost in the reduction, the convenience of data reduction to the decision maker is obvious.

Consider the basic steps of the scientific method. The scientific method is a highly formalized reasoning process containing the following six steps.

1. The event to be studied is defined.
2. Observations are made.
3. A hypothesis is postulated on the basis of a theory that seeks to explain the relationship among the observed variables.
4. An experiment is designed to test the hypothesis and validate the theory.
5. The experiment calls for measurements.
6. Results of the experiment validate or invalidate the theory.
    (a) If the hypothesis is accepted, the theory is admitted as a good explanation of the observed event. The explanation leads to generalizations and to laws as well as to predictions of future states.
    (b) If the hypothesis is rejected, a new theory and a new hypothesis must be formulated. Another iteration through the "hypothesis–experiment–test cycle" is carried out [p. 181; this quote and subsequent quotes cited to van Gigch (1978) are reprinted from *Applied General Systems Theory*, 2nd Ed., by J. P. van Gigch, copyright © 1978 by John P. van Gigch, reprinted by permission of Harper & Row Publishers, Inc.].

The purpose of this step-like procedure is to make a decision, to come to some conclusion about the way the world either is or should be to maximize some related contingencies.

We will not pursue here which is the chicken and which is the egg. To be sure, social science research methodology found a comfortable home within the decision-oriented perspective on evaluation and at the same time added fuel to the flames of the movement. Rather, from the perspective of evaluation, the remainder of this chapter examines more closely some of the methodological inadequacies and improprieties of social science methodology generally, and psychometrics and statistical inference particularly.

THE INADEQUACY OF THE TRADITIONAL
SCIENCE PARADIGM IN EVALUATION

The empiricist movement of the eighteenth and nineteenth centuries heavily shaped what was to become the social sciences (and subsequently, evaluation). The scientific method, which emanated from the work and writings of Descartes, proved an invaluable heuristic that did much to make possible the many startling discoveries of the day. It is not surprising that when human behavior came within the purview of scientists, they applied their tried-and-true methods for understanding and controlling the world. And the social sciences have been making strides ever since.

But the pace of scientific advancement in the social sciences has not matched that in the physical sciences. Given the disparity between the contributions and impact of these two metafields of inquiry, one wonders if the methodology borrowed from the physical sciences is particularly well-suited to the problems encountered in the social sciences. This is certainly not a novel contention by the authors, as the following quote from Koch (1969) indicates.

> A century and quarter ago, John Stuart Mill argued that the backward state of the social sciences could be remedied only by applying to them the methods of physical science, "duly extended and generalized." His strategy has been applied in billions of man-hours of research, ardent theoretical thinking, scholarship, writing, planning and administration in hundreds of laboratories by thousands of investigators. It has generated a vast literature and attracted generous support. Federal resources alone provided $326 million for "social science" research in 1967.[1] The test of the Millian hypothesis has not been a sleazy one. In my estimation, the hypothesis has been fully disconfirmed [p. 66; this quote is reprinted from *Psychology Today* Magazine, copyright © 1969, Ziff-Davis Publishing Company].

In a similar vein Koch (1969) states,

> In many fields close to the heart of the psychological studies, such concepts as *law, experiment, measurement, variables, control, and theory* do not behave as their homonyms do in the established sciences. Thus, the term *science* cannot properly be applied to perception, cognition, motivation, learning, social psychology, psychopathology, personology, esthetics, the study of creativity or the empirical study of phenomena relevant to the domains of the extant humanities. To persist in applying this highly charged metaphor is to shackle these fields with highly unrealistic expectations; the inevitable heuristic effect is the enaction of imitation science [p. 67].

In the authors' opinion, the restraints and limitations of analytical-mechanistic approaches borrowed from the physical sciences have their most obfuscating effect when we use them to determine whether or not our social science interventions are doing what they are supposed to do. Thus, it is in the field of program evaluation that the practice of borrowing hard science methodology has its most detrimental effect.

[1]More recently, Carol Weiss (1977a) cites estimates of federal funding of social science research from $493 million to over $700 million in 1976.

Van Gigch (1978) points out several features of the approaches associated with the scientific method that do not satisfactorily account for the characteristics of biological, behavioral, and social systems. Major limitations of analytical–mechanistic approaches especially relevant to program evaluation are listed below.

1. They could not completely explain phenomena such as organization, maintenance, regulation, and other biological processes that are characteristic of living systems.
2. The analytical method was not suitable to the study of systems that had to be treated holistically; the existence of irreducible wholes made the decomposition into component parts either meaningless or impossible. The properties of the total system of this kind could not be inferred from the properties of the parts, an important assumption of the analytical–mechanistic approach.
3. The mechanistic theories were not designed to deal with systems of organized complexity that displayed complex structures coupled with strong interactions.
4. The goal-seeking behavior of living systems, an important characteristic of open systems, required a theoretical foundation that could not be provided by the teleological explanations of old or by the cause–effect relationships of theoretical physics [p. 48].

Although the shortcomings of the scientific method as applied to the social sciences provide ample food for thought to the social science researcher, we are more concerned with the applicability of these procedures to evaluative inquiry. As Cronbach and his colleagues (1980) have pointed out, "Scientific quality is not the principal standard; an evaluation should aim to be comprehensible, correct and complete, and credible to partisans on all sides [p. 11]." Several major potential dangers hidden within the seemingly harmless application of the scientific method are discussed below. We believe these caveats to be as important as those cited by Stufflebeam et al. (1971). The limitations cited by those authors represented primarily procedural artifacts of experimental design inappropriate or inconvenient for purposes of program evaluation. The shortcomings cited below represent assumptions or inferential leaps imbedded within the role functioning of evaluators. Evaluators who have an implicit decision-oriented notion of their jobs and who rely extensively on experimental designs (and their extensions) to perform the tasks expected of them are most likely to fall victim to these inferential leaps. The potential illegitimacy of these inferences can seriously undermine the reliability, validity, and relevance of evaluation results which in turn, weaken the link between the evaluator and decision makers.

PROBLEMS WITH MEASUREMENT AND STATISTICAL POWER

The development of evaluation has been closely tied to advancements in the field of psychometrics. Some writers have even equated evaluation to measurement (cf. Thorndike and Hagen, 1969). Equating the two has the advantage of building directly on the scientific measurement movement with its attendant objectivity and reliability. Furthermore, measurement instruments yield data that are mathematically and statistically manipulatable, facilitating the establishment of norms and standards and fostering the use of experimental designs in evaluation contexts. The disadvantage of this view of evaluation is that it is totally dependent on the development, administration, scoring, and interpretation of measurement instruments (tests, questionnaires, attitude scales, etc.) which take time to develop and are relatively expensive. This approach also obscures judgment and judgment criteria. Scores become entities unto themselves while concepts behind the scores tend to be obfuscated. A final disadvantage, and perhaps the most important, is that variables that do not lend themselves readily to measurement are often eliminated or ignored.

Robert Boruch and Gomez (1977) offer an insightful view of measurement sensitivity and statistical power in impact (outcome) evaluations. They put forth the view that traditional experimental designs and measures are insensitive (lack statistical power) in the field applications required for evaluation. Rather than eschewing the use of experimental design in evaluation, however, these writers argue for augmenting orthodox measurement and design models to better reflect the circumstances encountered in evaluation contexts. They believe that insensitive evaluation designs are responsible for the fact that the vast majority of evaluative studies yield nonsignificant differences. Boruch and Gomez posit what they call "a small theory of measurement in field evaluation" to explain this insensitivity. This theory puts forth the notion that, owing partly to irrelevance of the outcome measure taken and partly to a lack of fidelity of the treatment administered in the field, the power of a design in the field setting is much lower than the same design implemented under laboratory conditions.

If one assumes that the response measure used is well matched to the elements and intents of the treatment and maintains its validity in the field (we will call this the response validity assumption), and assumes that the treatment implemented is exactly the treatment intended (we will call this the treatment fidelity assumption), then the relationship between treatment (t) and response (r) is given by Equation (2.1).

$$r = t + e \qquad (2.1)$$
$$e \sim (0, \sigma^2)$$

Of course our measure (y) of response (r) will not be perfectly reliable; thus:

$$y = r + \xi \tag{2.2}$$
$$\xi \sim (0, \sigma_\xi^2),$$

where ξ represents the random variation in the response measure (reliability).

The assumption of response validity. When the assumption of response validity is not met, however, Equations (2.1) and (2.2) no longer hold. Experienced evaluators can easily recall situations where the match between treatment components and response measures was less than ideal. Consider the common practice in educational evaluation settings of utilizing standardized achievement tests to assess the functioning of both the larger school system and specific programs within the system. The degree of overlap between intended program effects and the specific behavioral responses constituting a response measure will, of course, vary from program to program, and within a program will vary from measure to measure. Boruch and Gomez propose an *index of overlap or validity* that varies between 0.00 (indicating no overlap between program intents and the response measure utilized) and 1.00 (indicating a perfect match between program intent and what is being tapped by the measure in use). It might be that a particular achievement test overlaps .75 with the general cognitive goals of a school system but overlaps only .40 with the goals of a particular compensatory education program for some special population.

Thus, while Equation (2.2) depicts a response in the laboratory, Equation (2.3) represents the response variable when the response validity assumption is not made:

$$Y = K(r) + \phi + \xi, \tag{2.3}$$

where K indicates the degree of overlap between program intent and outcome measure ($0 \le K \le 1$) and ϕ represents irrelevant variability induced by the lack of overlap [$\phi \sim (0, \sigma_\phi^2)$]. Comparison of Equation (2.2) and Equation (2.3) illustrates algebraically the difference between the ideal response measure usually assumed in evaluation studies, and the model for a score when this assumption is not made.

The issue of response validity would certainly seem to warrant the attention of evaluators. Boruch and Gomez point out two sources of invalidity in the response measures used in evaluation. The first is the absence of a theoretical linkage between elements of treatment and elements of the response. The other is the "degradation of this nice measure in field settings [p. 413]." Given the nature of human service program formulation and funding by the federal government during the Great Society era and subsequently, theoretical linkages between program components and program *intents* were weak; the linkage with *measures* could be assumed to be weaker still. Certainly the vague, general, and ambiguous goals of most programs, usually adopted to achieve some political (or other) consensus, provide leeway for implementation and evaluation strategies in the field. Given vague objectives, the way these were operationalized also varied greatly, producing multiplicative sources of variation and mushrooming the variance of aggregate data. We shall have more to say about these more generic program linkage or assumption problems subsequently. The point to be made here is that the usual vagueness of program goals further inflates response invalidity. There is ample reason to believe that the assumption of response validity weakens the power of evaluation designs in both education and human service.

The assumption of treatment fidelity. Equation (2.1) assumes the treatment variable (t) to be ideally implemented. In classical experimental design the treatment is either presented or it is not and comparisons are made between these two states. In field applications, however, implementation of a program is not an all-or-none phenomenon. The treatment program-as-intended is implemented at sites in the field with various degrees of fidelity. Like response validity of an outcome measure, treatment fidelity can be expressed as a percentage of the ideal. Thus, treatment as implemented in the field (T) can be expressed as in Equation (2.4):

$$T = M(t) + \delta, \qquad (2.4)$$

where M indicates the degree of overlap between the program-as-intended and the program-as-implemented ($0 \leq M \leq 1$) and δ represents the variability in the actual level of the program imposed on or received by each program participant due to individual differences [$\delta \sim (0, \sigma_\delta^2)$].

Of course in programs that are implemented at several sites, there will be some differences between sites on this "*index of fidelity.*" In very large programs with many application sites, this problem is accentuated. Like the goals of these large-scale programs, the guidelines describing the structural integrity and composition of the program are often vague, ambiguous, and even if they do not defer to the implementation whims of local project administrators, these local preferences are likely to predominate. Often these "programs" are not actually programs at all but rather are policies handed down by the governmental bureaucracy. In these cases local site implementers are often given wide discretion in bringing a "program" into operation. The degree of overlap between the structural composition of any particular program and some standard or ideal could be measured in some way but seldom is. Instead, the assumption is made that the program being evaluated is precisely the program intended.

The reader should note that "no-treatment" strategies implemented in the field for purposes of statistical control could be subjected to similar analyses. Seldom is any effort spent ascertaining whether the control group(s) received any treatment or not. It is easy to imagine how field staffs might not withhold potentially beneficial services from a client or program participant at the whim of an evaluator. This would induce still another source of variation into evaluation design models utilizing no-treatment controls, further reducing power.

The problems of program implementation have not been totally ignored. Certainly there is a growing literature on the issue of implementation alone. The following quote from a text on social program implementation edited by Williams and Elmore (1976) illustrates the thrust of this movement: "The most pressing implementation problem is that of moving from a decision to operations in such a way that what is done bears a reasonable resemblance to the decision and functions adequately in its institutional environment [p. 4]." Similarly, Cronbach et al. (1980) speak of "treatment realizations" and offer some symbolic language to represent the idea that any ideal model of treatment will be represented by a domain of treatment realizations in the field.

What is the case when we do not make the assumptions of response validity and treatment fidelity? By substituting in Equation (2.2) (the equation for a predicted score in the laboratory) the terms in Equations (2.3) and (2.4) (which replace assumptions of response validity and treatment fidelity), we obtain the model for a response in the field. This substitution is made in Equation (2.5):

$$Y = KMt + K\delta + Ke + \phi + \xi, \tag{2.5}$$

where K indicates the degree of overlap between response measure and program intent ($0 \leq K \leq 1$); M indicates the index of structural integrity or fidelity of the program as implemented ($0 \leq M \leq 1$); t indicates treatment; δ indicates variability in the treatment received by each individual; e indicates unspecified, possibly unknowable influences; ϕ indicates irrelevant variation in response variable induced by lack of overlap between response measure and program intent (i.e., invalidity); and ξ indicates measurement error, usually called the reliability of a measure.

Comparing Equation (2.2) with Equation (2.5) illustrates the considerable differences that exist between field and laboratory applications of what is generally assumed to be the same experimental design. Boruch and Gomez cite the hypothetical example of a program for 40 participants showing a half standard deviation treatment effect in the laboratory. Setting the significance level at .01, the power of the statistical test will be about .92. The chances are about 9 in 10 that a difference will be detected. Substituting the augmented model for a score given in Equation (2.5) changes the power of the design. Assuming 75% relevance of the response measure, a treatment fidelity of .75 and outcome measure reliability of .80, it can be shown algebraically that the power of this design drops to about .30. Here the chances are 3 in 10 that the half standard deviation effect will be detected in the field. This drop in power is considerable.

Disregarding for the moment other limitations of decision-oriented use of research designs in evaluation, one questions a practice that repeatedly yields no significant differences. When the bottom line of a traditional evaluation study is a significant difference between treatments, this is thought to be well and good. You may have something to say. When significant differences are not found, evaluators are put in the role of shrugging their shoulders and saying "sorry guys." Program planners, developers, and implementers (usually with decision making authority) scratch their heads, wonder what went wrong or conclude that the program does work and decide the evaluator was incompetent and is not worth consulting in future decision making about this and perhaps other issues. If a finding of no significant difference is the lion's share of information provided by an evaluator, dismissing evaluative input is understandable. There is very little that can be done with that information. It leaves decision makers and policy shapers having to use criteria other than effectiveness as the basis for decisions about programs. Furthermore, when the situation exists where program costs are not a major issue, as was the case during the Great Society era of program expansion, indeed the only criteria are those of special interest groups and political feasibility.

Boruch and Gomez's small theory of measurement in field evaluation provides an insightful view of problems associated with experimental design in evaluation. We agree with Boruch and Gomez's conclusion that the small theory is an overly simplistic approximation of a theory of measurement in the field, but their work represents a needed first step. The theory illustrates the fact that the effects of response invalidity and treatment infidelity interact multiplicatively rather than additively to reduce power in evaluation studies.

Boruch and Gomez (1977) go on to cite considerations that help pinpoint fidelity of response and treatment but admit that at best there are "fragments of methodology that may be helpful" here. How do you measure, for example, a particular evaluation design? Or how does one quantify the fidelity of the treatment program being evaluated? How is an index of fidelity calculated? While their thinking on this issue is insightful, their recommendations are more descriptive of present problems than they are suggestive of solutions or new methodology.

Boruch and Gomez do, however, provide a rich conceptual framework for thinking about field measurement and design problems and at a minimum suggest areas and questions within the evaluative context that beg for the development of alternative methodology. We should also mention the recent work on this topic by Cronbach *et al.* (1980). These writers consider these same problems in a more qualitative perspective, pointing out the differences between the field investigator's desire for an elegant experimental design and the concerns of program implementors in local sites. The remaining chapters of this book will be concerned with suggesting and explicating evaluation perspectives other than experimental designs to address these issues.

Before moving beyond issues of measurement and statistical design we would like to point out some decisions implicit in the statistical approach emphasized by Boruch and Gomez, among others. These comments apply regardless of the augmentations, extentions, or transformations one may apply to experimental designs to increase their adequacy in the field. The first is the decision about data. To measure and statistically analyze, one must decide (assume) that the data are meaningful; that they are reflective of the true state of affairs. Once the decision to measure has been made, one must decide about an experimental design. We want to choose a design we believe will be sensitive to the "answer" we believe to be in the data. Once statistical results have been generated, one must decide how to interpret the data. The potential for bias and faulty assumptions at each of these decision points can further invalidate evaluation results.

GENERIC INFERENCES IN PROGRAM EVALUATION

The term *inference* is an oft-used term in the language of experimental design and statistics. *Parametric statistics* refers to little more than collecting data on a relatively small number of cases and drawing inferences or generalizations from analysis of the data about some larger population of which the analyzed cases are considered representative. As the foregoing discussion illustrates, however, unwarranted assumptions are often made in the process of measurement and analysis that render statistical inferences less potent, if not impotent.

Unfortunately this is not the only inferential leap of faith made in the process of generating and analyzing data for purposes of program evaluation. Utilization of experimental designs involves making concurrent, parallel assumptions regarding the context in which a program operates. While these assumptions overlap with those associated with experimental design, they independently and simultaneously contribute to a lack of clarity, credibility, validity, and utility of evaluation results. These more universal assumptions, which are often made regardless of the evaluation approach utilized, are examined in turn below.

Legitimacy of program objectives. In our discussion of response validity of evaluation outcome measures, the primary issue was whether the response measure tapped precisely the domain of behavior change intended by the program. In this section we put forth the notion that these intents, usually called goals or objectives, are themselves assumed to match the structure and content of the program and to follow logically from some undergirding theory or rationale. The assumption that the response measure is a valid measure of program objectives carries with it the parallel assumption that program objectives themselves can logically be expected to accrue, given the nature of the program and the philosophy behind it.

In discussing response validity, Boruch and Gomez (1977) point out the need for a "rhetorical program model" that depicts assumed interrelationships between components and objectives. They consider such a model to constitute a rationale or "minitheory" for the link between treatment and response. However, in some cases, program objectives have become entities unto themselves and are taken as givens. The rationale for how the program is to attain its objectives may never be specified. For the evaluator, the task is often defined as one of finding a response measure congruent with a particular objective. If the objective could not logically be expected to occur as a result of the program, then the power of any statistical design employed with that measure is further undermined. The probability of detecting a true program effect, if indeed there is one, grows even smaller.

In many cases, this situation results when the goals of programs are global, vague, and overly ambitious. Broskowski and Driscoll (1978) point out that this set of circumstances often occurs in federally funded human service programs where altruistic, ambiguous goals are utilized as integrative organizational mechanisms. Goal ambiguity fosters consensus about the purpose and function of the program. Highly ambitious goals provide incentives for performance. In this way vague, ambitious goals become the undergirding rationale for a program.

In discussing accountability and goal clarity, Demone, Schulberg, and Broskowski (1978) reach similar conclusions. Citing Mogulof (1973) and Horst et al. (1974), they state that the vagueness and generalities required to achieve Congressional consensus result in federal programs lacking measurable goal attainment criteria. They go on to state that "Program interventions are usually vaguely defined, not linked to measurable outcomes or ultimate impact by testable assumptions, and then operated by "pseudomanagers" who lack power or incentives to control resources or implement evaluation findings [p. 34]."

The legitimacy of program objectives is an active but often ignored issue in many federally funded programs, indeed in any program where global, altruistic outcomes have been adopted as program objectives. On the surface, such global objectives as human equality, equal employment, educational opportunity, financial assistance to the needy, equal rights, and access to medical treatment are hard to argue against as appropriate program goals;

they are so humanitarian. Only a truly villainous person would doubt the legitimacy of these goals.

The problem lies in the fact that these desired ends represent more the policy objectives of political units or administrative structures in power than they do the objectives of specific programs. Once policy objectives have been determined, programs are created (or existing programs modified) to bring these objectives to fruition. As implied earlier, program design and implementation generally receive less attention than do policy objectives for each program emanating from a specific policy. The burden is on local program personnel to operationalize the program and its objectives. Given this freedom, program administrators may operationalize programs and objectives as they see fit, creating an environment where the evaluator may be forced or co-opted into a restricted definition of his or her role. In this way, vague objectives and guidelines contribute to the growth and nurturance of decision-oriented evaluation practices.

The critical nature of the match between program theory and components on the one hand and objectives on the other certainly is not new in evaluation (cf. Riecken, 1952). Stake (1967), for example, proposed a preliminary phase of evaluation in which the *logical* contingencies existing between intended antecedents, transactions, and outcomes are examined before actual data from the field are collected and analyzed (see p. 8, this volume).

This "front-end analysis" (Harless, 1973) would uncover logical flaws in the program that go undetected by experimental designs, regardless of their power or robustness. It seems reasonable to propose that after an evaluator has been introduced to a program, his or her first question should be "Is there reason to expect this program will work?" After some familiarity with the program, the evaluator may decide it is easier to enumerate reasons the program will not work!

Methodologies for laying out logical contingencies are not well formalized, although some do exist. Wholey, Nay, Scanlon, and Schmidt (1975), for example, propose the use of a model depicting assumed linkages between program activities and objectives. This model then serves as a basis for determining the evaluability of a program. Program modeling represents a new arena of evaluation expertise in this larger, systems view of programs and the role of program evaluators. We shall have more to say about program modeling and other more systems-oriented evaluation activities subsequently. It should suffice to say here that the idea of examining the assumed relationship between a program and its objectives is not a new one in evaluation. However, methodology for conducting such inquiry has traditionally been lacking.

Representativeness of program objectives. Another popular assumption is that program impact occurs only in those domains of behavior tapped by outcome measures. With the use of experimental designs, measures of variables tend to become real entities, sometimes to the point where "scores" actually become the variable of interest, while the attribute supposedly being measured becomes obfuscated. Another problem with this practice is that if a variable is not built into an evaluation design, the effect of the program on that variable will go undetected.

Given the myriad aspects of behavior any particular human service or educational program might affect and the relatively small number of outcome variables usually tapped by an experimental design, it seems safe to conclude that we make judgments about programs with much less than complete data. Furthermore, as has been pointed out previously, the data selected as the focus of evaluation may be advantageously biased in favor of some individual, group or interest. Thus, if one were really interested in the effects of a program one would look beyond program objectives to determine if any unintended or unavoidable side effects, positive or negative, simultaneously accrue as the program strives to attain its objectives.

These ideas are borrowed in part from Scriven (1967, 1974) and his notions of goal-free evaluation and the place of values in evaluation. We shall examine this value-oriented perspective on evaluation more thoroughly in the next chapter. Here we would like merely to point out that the assumption that program objectives are representative of all program consequences will sometimes constitute a logical flaw that renders the results of a particular analysis misleading.

Appropriateness of program objectives. A major thrust of the value-oriented approach to evaluation (Borich and Jemelka, 1981; Jemelka and Borich, 1979) concerns the match between program objectives and the wants, values, and needs of those for whom the program is intended. Even though all the previous caveats might not apply to a program, if program results are not desired, valued, and needed, who cares that the statistical analysis shows the program to be very effective in producing the stated objectives?

This realization, more than anything else, has forced evaluators out of a strictly decision-oriented stance and into the arena of providing information for program planning and development, primarily through the developing methodologies of needs and policy assessment. We shall introduce these topics more thoroughly in subsequent chapters. The point to be made here is that one should be aware of where the program objectives came from. Were they developed in response to a careful study of the needs of some particular client population, or do they merely reflect political reality or the personal interests of some individual group or constituency? Experienced evaluators know well that assuming the appropriateness of program objectives can at times be quite tenuous.

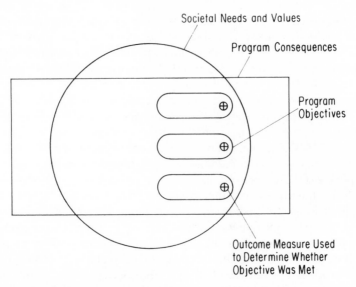

Figure 2.2. **The many domains of program evaluation: societal needs, program consequences, program objectives, and outcome measures.**

The foregoing discussions about the legitimacy, representativeness, and appropriateness of program objectives illustrate some of the ways in which the accuracy, comprehensiveness, and meaningfulness of statistical evaluation results may be impugned. These assumptions, often made within a decision-oriented perspective, are illustrated in Figure 2.2.

Figure 2.2 attempts to show that even if all the assumptions underlying the use of a decision-oriented strategy were met and the evaluation were meticulously carried out in the decision-oriented tradition, the fidelity, representativeness, and appropriateness of program objectives could not be guaranteed by the decision-oriented approach alone. There would still be a chance that a particular outcome measure may not be faithful to program objectives, objectives focused upon may not be representative of all outcomes produced by the program, and objectives may not be realistic or may be a poor reflection of whether the program produces ends desired or valued by society. There is nothing in this methodology to prevent these "lacks" from evading the myopic view of the decision-oriented evaluator. Moreover, because each principle actor in the program has a different viewpoint concerning program objectives, the issue is further clouded. Although decision-oriented program evaluation may respond satisfactorily to questions concerning the achievement of given program objectives, it is less capable of systematically determining the legitimacy of program outcomes, and almost completely negligent in examining the representativeness and appropriateness of these outcomes.

New Directions for Evaluation

Given the foregoing discussion of the evolution of decision-oriented evaluation and the methodological and logical flaws that may exist within it, one wonders why this perspective has been so prevalent in the work of evaluators. Of the large number of texts on evaluation, many present a decision-oriented view with primary emphasis on the use of traditional experimental designs in evaluation settings. For example, Rossi, Freeman, and Wright (1979) devote considerable space to discussions of randomized, controlled experiments, as do Riecken and Boruch (1974), viewing this as the optimal approach to outcome evaluation. Why has the call for new and different methodology, inherent in the work of so many influential writers in the field of evaluation in the last two decades, fallen on deaf ears? Indeed the terms *applied social research* and *evaluation,* despite a recent broadening of evaluation methodology, are often used interchangeably in a large portion of the literature.

Notable exceptions are represented by the work of Britan (1978), Cronbach et al. (1980), Guba (1978), Guba and Lincoln (1981), Parlett and Hamilton (1976), Stake (1975b), and Patton (1980). These authors have pioneered in the use of naturalistic inquiry and alternative conceptual frameworks for evaluation. Guba (1969), in particular, should be cited for following the suggestions of his own caveat put forth nearly a decade earlier:

> When the evidence produced by any scientific concept or technique continually fails to affirm experimental observation and theory arising from that observation, the technique may itself appropriately be called into question. It shall be the burden of my remarks that evaluation as we know it has failed, and that the world of evaluation does indeed require reshaping [p. 30].

After reviewing 181 evaluation studies utilizing experimental designs of a "high methodological caliber," Mann (1972) came to similar conclusions. He found that behavior change was demonstrated in about 45% of these studies but findings were unrelated to the type of intervention being evaluated (e.g., individual versus group treatment), the content area of the program (e.g., education or community mental health), the change criteria being used (e.g., follow-up questionnaires versus behavioral observations), or the methodological quality of the experimental design and statistical procedures utilized. Mann concludes ". . . that evaluative research represents a scientific blind alley. It has failed to validate itself in practice and the sooner its failure is accepted and recognized, the easier will be a transition to another approach to the same problem [pp. 278-279]."

The primary thrust of the decision-oriented approach to evaluation as theoretically formulated by Alkin (1969), Stufflebeam et al. (1971), and others was to maximize the utilization of evaluation results. Decision-oriented evaluation, as we perceive it to exist in the real world of programs, has fallen short of the intentions of its sage founders. Done in by the subservient administrative role usually afforded evaluators and their own blind adherence to the pursuit of the fruits of a keen sense of experimental design, the results of their labors have been poorly utilized.

Utilization has become a major issue in program evaluation. Weiss (1972a) provides a thorough discussion of major issues moderating the acceptance of evaluation procedures and results, including:

1. Conflicting perceptions of the purposes that evaluation is expected to serve
2. The place of evaluation within the organizational structure
3. Relationships between evaluators and practitioners
4. Characteristics of the program that hamper evaluation
5. Timing
6. Use of evaluation results [p. 3]

In another article Weiss (1972b) suggests several activities that might increase the use of evaluation results. Among these are an explication of the underlying rationale for the program and the specification of a process model of a program detailing linkages between input, process, and output. We saw earlier that these were major issues, that, if ignored, may result in an undermining of the power and validity of experimental designs. In a similar vein Attkisson, Brown, and Hargreaves (1978) have questioned whether the field of evaluation is capable of responding to the demands being placed upon it. They cite five reasons why evaluation is disregarded by human service administrators. These are (a) the lack of validity caused by the difficulties of doing "clean" research in applied settings, (b) the failure to design evaluations having managerial utility, (c) the rejection of negative evaluation results, (d) the unwillingness or inability of management staffs to incorporate evaluative input, and (e) the complexity of administrative decision making.

We view most of the utilization problems cited by Weiss and by Attkisson, Brown, and Hargreaves as symptomatic of decision-oriented evaluation. It is ironic that the decision-oriented perspective, which was an attempt to foster utilization in theory, has resulted in such poor utilization of evaluation results in practice.

SOME SUGGESTED DIRECTIONS FOR GROWTH

Evaluation has been slow to follow up on suggested areas from which developing evaluation methodology could be gleaned and to integrate these methodologies with current practice. As early as 1960, Etzioni suggested a systems model as opposed to a goal model for evaluating programs. His argument was that organizations, viewed as systems, engage in a wide variety of activities besides goal achievement. Thus, an evaluation focusing exclusively on goal attainment will miss other areas of program effectiveness or failure. The careful reader will note the similarities between Etzioni's (1960) approach and Scriven's (1973) goal-free evaluation.

Even the methodological traditionalist E. A. Suchman, whose classic 1967 text did much to tie the field of evaluation to the scientific research model, calls for "systems-oriented evaluation" and the utilization of operations research in evaluation practice. Suchman was aware of the open systems concepts and dynamic process models envisioned by advanced systems thinkers of the day, primarily the seminal work of Buckley (1968). Although Suchman's systems-oriented evaluation is a more limited view of the systems approach to evaluation than we present in this volume, it is undeniable that like many others he was aware of the need for new paradigms in the field of evaluation.

Stufflebeam *et al.* (1971) also emphasized the need for a systems paradigm in their classic text on decision-oriented evaluation. While it was unclear at the time just exactly how systems concepts could benefit the work of evaluation, there was a belief among these writers that systems theory concepts had much to offer the evaluator in terms of understanding the structure of programs and their operation within a larger context.

Robert Weiss and his associates (Weiss and Rein, 1972) proposed that the primary question usually asked in evaluation, "Does a program work?," be supplanted with the question, "When such a program is introduced, what then happens?" They suggested adopting alternative methodologies drawing from conceptual frameworks other than experimental design, including systems theory, historical perspective, and political interaction models. Their framework suggests a nonquantitative, process-oriented approach which functions to direct attention "to the sorts of events which should be recorded in data gathering, to the questions which must be answered in the analysis, and to the kinds of connections which should be demonstrated in the report [p. 243]." Weiss and Rein's potentially seminal work in the late 1960s and the early 1970s failed, however, to promote interest and activity in the search for alternative methodological frameworks for evaluation.

The foregoing brief review illustrates that the need for new methodology and alternative conceptualizations and frameworks for evaluation is not a new phenomenon. Certainly most of what has been put forth in this chapter can be found in the writing of the clear thinkers in the field of evaluation over the last quarter of a century. Riecken's (1953) memorandum on program evaluation is one such example. Obviously, the ideas have been kicking around for a long time. These ideas have done little, however, to foster the utilization of new conceptualizations and methodology for evaluation. While a climate for change can now be noted, until very recently the field of evaluation has plodded down its well-traveled (and worn) path. We shall speculate on some reasons for this below.

REASONS FOR RESISTANCE.

Carol Weiss (1972c, 1977b) has provided a thorough analysis of the juxtaposition of program evaluation and the ongoing political context in which it operates. She points out that because evaluation studies often have political consequences, they will be attacked by those whose ideas and values are unsupported by the data. The scientific respectability of close adherence to experimental design limits the evaluator's liability to criticisms of the study, particularly its methodology.

To employ some less well-established methodology is to leave oneself open to vicious methodological criticism. Cut off from administrative decision making and power, few evaluators have been willing to risk such methodological innovations. Weiss's (1972c) description of this reticence follows.

> Systems approaches and process-oriented qualitative analyses, for example, are being applied to large-scale programs. But none of these departures, not even the quasi-experiment, has attained the legitimacy of experimental design. When methodology is subject to attack, evaluators are wary of the untried tack. Many apparently prefer to stick unimaginatively to the book, rather than risk the penalties of pioneering [p. 330].

Weiss also addresses the autonomy of the evaluation effort, citing the potential practices and activities that might undermine an evaluators' independence. We believe the issue of autonomy to be central to the role played by evaluation in programs. Lack of autonomy has in many cases greatly reduced the role played by evaluators in program planning, development, evaluation, and decision making.

Summary

The foregoing review reveals the failure of traditional evaluative inquiry to promote an understanding of the programs we create, implement and live with. The vast diversity of procedures and practices that have been accepted as "evaluation" illustrates our haphazard commitment to knowing which programs are capable of obtaining what outcomes at what costs. Although the public clamors for accountability and each succeeding generation of politicians vows allegiance to this principle, only in a few instances has program evaluation activity produced a credible and multipartisan evaluation report.

In some ways a major portion of the history of many fields reveals an unknowingly concerted effort to avoid accountability. Educational systems, mental health care delivery systems, and public welfare systems are but a few examples of systems that have become monsterously large, unmanageable, and unaccountable, yet seemingly absolutely necessary for the continued functioning of our society. How did this state of affairs eventuate? One of the primary reasons may be in the way we have attempted to deal with complexity in our society. This book presupposes this lack of accountability to be the result of the accumulation of inadequate short-term solutions, inadequate justification for funding, and avoidance of personal responsibility by key individuals within systems.

Short-term, temporary solutions to larger-scale problems, poorly thought out justification for federal funds (and granting funds despite poor proposals), and avoidance of personal responsibility have become our modus operandi for dealing with increasing complexity in our society. Implementing cosmetic changes for the sake of appearances, responding to confusion by spending more money and hiding one's level of incompetence have apparently become part of the American ethos. As long as the system continues to choose to behave this way, little change can be expected to occur. However, such events as the legislated Stull bill and the passage of Proposition 13 in California as well as more recent cutbacks in public funding are symptoms of a society demanding more accountability of its programs.

The question remains as to how we achieve accountability. Given that effort and expertise expended towards this end have not produced overwhelming results, one questions whether the state-of-the-art of our evaluative methodology is currently mature enough to respond to the expectations being placed upon it. The question is, how do we shift from an unaccountable to an accountable way of providing services?

In the following chapters we articulate some broader perspectives on evaluation than those found in the decision-oriented literature and practice of evaluation. The next chapter elucidates the idea of value-oriented evaluation, which dramatically expands the turf of the evaluator. This value-oriented notion of evaluation, articulated by Scriven (1967, 1974), Cooley and Lohnes (1976), and Jemelka and Borich (1979), owes its philosophical foundation to early writings of educator John Dewey (1922, 1939). The value-oriented approach represents a strong reaction to the evolving practices of decision-oriented evaluation and suggests some new tools and concepts for the evaluation of programs, including unobtrusive measurements, naturalistic inquiry, and needs assessment methodology. Then, in Chapter 4, we introduce some systems-oriented notions and concepts. From Chapter 5 to the conclusion of this book we increasingly focus and define our own perspective on evaluation, drawing heavily upon the concepts of value and systems-oriented evaluation.

Before concluding this chapter, however, we want to convey to the reader who may have been put off by the notions put forth in this chapter that we do not view the seminal work of the founders of the theoretical notions of decision-oriented evaluation to be weak, flawed, or damaging to the field. These individuals have been among the most enlightened thinkers about the needed futures of evaluation. They have not been the culprits. Nor are we saying that evaluation should be value-oriented or systems-oriented as opposed to decision-oriented. These are descriptive conceptual "handles" rather than actual entities. Real life evaluation contexts will bear semblances of all three. Furthermore, there are reasons why decision-oriented practice came to predominate and these reasons have not disappeared.

What we are saying is that regardless of the context in which evaluators find themselves, they must let their thinking be guided by something beyond a goal attainment orientation, *exclusive* reliance on experimental design and related statistics, and the immediate context in which they ply their trade. The following chapters provide some conceptualizations, ways of thinking, methodological tools, and heuristics that we believe are consistent with the reform of program evaluation hailed by Cronbach and his colleagues (1980):

> For all the recent progress, the battle for reform has barely been joined. Many of the most recent textbooks in evaluation still take the stereotypic line. Summative appraisal that tells the decision maker which alternative action to take is held out as the aim of real evaluation. And the draft guidelines for federal evaluations that the GAO has circulated, while accepting the importance of fitting the evaluation plan to the political and administration context, nonetheless place heavy emphasis on controlled experiments to access impact. . . [pp. 74–75].

3
Value-Oriented Conceptualizations of Evaluation

Some authors in the field of evaluation have taken exception to the notion of decision-oriented evaluation and its implications for the conduct of evaluative research. The primary criticism against decision-oriented evaluation is that it removes the evaluator from the decision-making process. The role of the evaluator is merely to provide the decision maker with meaningful information; the decision maker makes the actual judgment of value or merit.

Value-oriented conceptualizations of evaluation, on the contrary, view evaluation as a shared function. The value-oriented approach stresses the value judgments made in evaluating programs, and furthermore describes the act of judging merit or worth as central to the role of the evaluator. Worthen and Sanders (1973), for example, define evaluation as "the determination of the worth of a thing [p. 19]." In addition, Scriven (1967) considers the evaluator who does not participate in the decision-making process as having abrogated his role. On the other hand, Stufflebeam et al. (1971) argue that by participating in decision making, the evaluator loses his or her objectivity and, hence, his or her utility. Similarly, Cronbach et al. (1980, pp. 154–155) argue that such judgments fall outside the realm of the evaluator's responsibility. Differences in these approaches are more than semantic, for they imply different evaluation activities.

Within the decision-oriented approach, the evaluator is dependent upon the decision maker for the way the decision context is to be defined and the values and criteria that are to be used to judge program success (these are usually termed program intents, goals, or purposes). However, Cooley and Lohnes (1976) and Apple (1974) point out that there is no evidence to suggest that the decision maker is any more capable than the evaluator of defining decision settings, alternatives, and values. Indeed there may be (and often are) social, institutional, and political presses on the decision maker which may lead him or her to opt for evaluation procedures that skirt or ignore key evaluation issues.

Apple (1974) makes the case that decision-oriented evaluation is a conservative practice not conducive to the acceptance of innovation but rather supportive of the status quo. His point is that the limits of the decision-oriented evaluator's work are circumscribed by the already developed program, and, therefore, the evaluator cannot deal with the issues, concerns, and goals that led up to the program and to which the program is supposed to be a response. Once the program is in place, the evaluator's role is to work with it regardless of whether it is the best means to the desired end.

Scriven (1973, 1974) argues that value judgments are a crucial part of all sciences, particularly methodological value judgments, and there is no reason to dismiss them in evaluation. He calls for goal-free evaluation, insisting that all aspects of a program should come under the scrutiny of the evaluator and that nothing should be taken as given from the client or agency soliciting evaluation expertise. The following illustrates his point:

> The goal-free evaluator is a hunter out alone and goes over the ground very carefully, looking for signs of any kind of game, setting speculative snares when in doubt. The goal-based evaluator, given a map that, supposedly, shows the main game trails, finds it hard to work quite so hard in the rest of the jungle [1973, p. 327].

Scriven argues that although knowledge of goals is necessary for effective planning and implementation, it is unnecessary in evaluation and may even blind the evaluator to important program effects.

Scriven (1973, 1974) and Apple (1974) also emphasize the social responsibility of the evaluator. Scriven offers the hypothetical example of a program aimed at increasing self-sufficiency. After some evaluative activity the evaluator discovers that in addition to fostering self-sufficiency, the program engenders contempt for the weak, sick, old, and congenitally deformed. Scriven contends that those findings should count against the program even if the program meets its announced and intended goal. The welfare of the consumer (usually in the case of education and human service, society as a whole) is considered a proper concern of the evaluator.

Apple (1974) puts forth a similar argument:

> The tendency in the face of the all-too-usual finding of "no significant difference" is to argue for better teacher training, for better instructional materials, for more sophisticated administrative systems designs and the like. However, it may well be that more basic questions must be asked, that even the obligatory nature of the

institution of schooling may need questioning, or that educators are asking the wrong kinds of questions.

For example, much low achievement on the part of many students could be attributable to a symbolic dismissal of school itself as a meaningful institution. These students may perceive schools as relatively unresponsive to human senti- ments. This is not to argue that schools should be done away with; to take such a position in a knowledge-based economy is somewhat unrealistic. It does signify, however, that educational problems are considerably more fundamental than educators may suppose, and it places responsibility on the individual educator to examine his or her own professional activity in a wider social and political con- text [pp. 28-29].

One implication of Apple's view for the evaluator has been elucidated by Becker (1974), a sociologist, whose comments foreshadow how the evaluator who fails to give deference to the status quo might be received by the decision maker:

For a great variety of reasons, well-known to sociologists, institutions are refrac- tory. They do not perform as society would like them to. Hospitals do not cure people; prisons do not rehabilitate prisoners; schools do not educate students. Since they are supposed to, officials develop ways of denying the failure of the institution to perform as it should and explaining those failures which cannot be hidden. An account of an institution's operation from the point of view of subor- dinates therefore casts doubt on the official line and may possibly expose it as a lie [p. 113].

Becker believes that any approach the evaluator might take is inherently value laden and will implicitly support either the subordinate (program par- ticipants') or superordinate (program manager's) point of view. Becker's comment also raises the possibility that due to the efforts of decision makers to protect the status quo or to limit changes to those that are congruent with the existing social, political, and organizational structure, evaluators may implicitly design evaluations that examine *only* the merits of these two points of view, avoiding all others. It is from the heated debate of conflicting points of view that evaluators most readily discern differing value positions. Therefore, any attempt to limit the conflict to the values espoused by pro- gram participants and management may also be thought of as suppressing the values of all others who might have a stake in the program. This is most likely to occur when the goals and objectives for a program must be taken as "givens" and the evaluation designed around them.

Cronbach et al. (1980) have also suggested that evaluations should be multipartisan. It is their view that a thorough evaluation makes the strongest case for each value position and that no single value perspective should dominate or be excluded. Thus the Cronbach group might disagree with

Becker's position that any evaluation perspective will necessarily favor any one value perspective over all others. These writers would argue, rather, that the value perspectives of all program stakeholders should be considered in the evaluation process.

Values in Evaluation

A cohesive value-oriented theoretical perspective on evaluation has been put forth by Cooley and Lohnes (1976). Their stance is based on the early work of John Dewey (1922, 1939) and borrows from Handy's work on the study of values in the behavioral sciences (Handy, 1969; Handy and Kurtz, 1964). While the propositions of Cooley and Lohnes's theory of evaluation are quite similar to and generally subsume those of Apple, Scriven, Worthen and Sanders, and others, they are put forth in a more direct fashion that has practical implications for the conduct of evaluation. They assert that the value statements inherent in programs can themselves "be analyzed into a set of propositions subjectable to empirical investigation and that failure to perform such analysis in evaluation studies is inexcusable [pp. 9-10]."

Cooley and Lohnes argue that the values that have guided educational practice have traditionally been determined by politics and custom, and that their validity has not been challenged by educational researchers. They find it curious that value propositions have evaded empirical scrutiny despite educational researchers' heavy emphasis on empiricism. They consider clear thinking about values in education to be essential because educational practice is generally influenced by the value attached to desired educational goals. The alternative to rational inquiry into values is the determination of values on the basis of power which places the educational enterprise "at the mercy of special interest groups who commend values favorable to themselves as universals [p. 10]." Although these authors write primarily from an educational program orientation, the problems and issues they address are relevant to human service program contexts as well.

An important premise in Cooley and Lohnes's argument is that values are subject to change. This premise reflects that values can be mistakenly viewed as absolutes only if they are considered out of context. When considered *in context,* values lend themselves to elucidation as propositions about real entities (matters of fact) and the error of ascribing to them absolute or universal properties is thus avoided. The task of the evaluator becomes one of ascertaining whether value propositions inherent in a program setting reflect only convention or tradition or whether they imply empirically testable relationships between means and ends.

Consider the hypothetical example in which an evaluator is called in to determine whether an inservice training program for teachers would increase the teacher's appreciation of the difficulties encountered by Spanish-speaking children in a predominantly English-speaking community. The foregoing discussion suggests that the evaluator should consider the context before proceeding. Did school administrators merely assume that a general inservice program would have this effect? Was pressure applied to administrators to improve teacher understanding of cultural differences? Was the program developed just because funds were available? Or was it developed because a survey of teachers, parents, and students indicated that such an inservice program would be beneficial? The latter possibility is desirable but seldom encountered.

The value judgment explicit in the above example is that teachers need to have a greater appreciation of the educational difficulties encountered by Spanish-speaking children. Implicit are judgments that the teachers presently are insensitive to these problems, that these students are being short-changed in their education, and that the administration is quite concerned over this state of affairs. Each of these value propositions may or may not be true but is capable of being empirically investigated.

Optimally, the need for such a program would be ascertained before it is developed and implemented. However, this planning activity is not always done. Evaluators are usually ignored in program planning, development, and often in implementation. The exclusion of evaluators greatly limits the evaluation expertise that could be brought to bear in the program setting. Evaluation has much to offer in terms of the "front end" work of program planning, and some inroads have been made in the area of needs assessment (Bell, Nguyen, Warheit, and Buhl, 1978; Siegel, Attkisson, and Carson, 1978; Kaufman and English, 1979; Neuber, 1980). This issue will be discussed subsequently, but let it suffice to say now that the notions of Dewey (1939), particularly as they are elucidated by Cooley and Lohnes (1976), provide theoretical justification for the involvement of evaluators early in program planning generally and in the measurement of values particularly.

Another significant contribution of Dewey's theory of evaluation to Cooley and Lohnes's perspective is the notion that there is no absolute distinction between means and ends. Any event or condition can be viewed as occupying space on a continuum such that it is simultaneously an end to those events and conditions that preceded it and a means to those that follow. While each event, condition, or program has its own "ends in view," these ends are meaningless if they are not congruent with higher level goals to which they are expected to contribute. Improved reading skills is an end that is useful only as a means to more comprehensive outcomes: promotion to the next grade, graduation from school, getting and holding a job, and ultimately, contributing in some meaningful way to the society in which the learner lives. Likewise, the usefulness of any drug abuse program can be properly conceptualized only in the context of higher level goals: physical and mental well-being, self sufficiency, regained productivity, and so on. Dewey (1922) makes the assertion that only when an end is conceptualized as a means is it fully understood, appreciated, or even obtainable.

At the program level the evaluator typically constructs a small means-end continuum by dividing outcomes into two classes: *enabling*—those that are prerequisite to the attainment of terminal outcomes—and *terminal*—those that are expected at program completion. Although this division can be viewed within the framework of a means-end continuum, it emphasizes goals or objectives that are internal to the program and hence offers no assurance that the program will be evaluated or even conceptualized as a means to subsequent ends. Provus (1971) and Stake (1967) extend this "internal view" of a program a step further by providing models of the evaluation process that contain "ultimate" outcomes and "personal and community wide" program effects, thereby linking the program to some larger framework. Still other authors (Hoban, 1977; Lasswell, 1948, 1960; Rucker, Arnspiger, and Brodbeck, 1969) have argued for ultimate outcomes that appeal to some higher order good chosen from among the values shared by our society such as affection, enlightenment, rectitude, respect, skill, power, wealth, and well-being. However, taking too external a view of the program might result in goals that, for all intents and purposes, cannot be operationalized or are unattainable within any specific program setting.

Delineation of the means-end continuum can be neither too broad nor too narrow and must occur within a particular context. It would be a mistake for the evaluator to conceptualize as ends abstract, unattainable entities that are external to the program, just as it would be shortsighted to view program objectives as the terminal point in the means-end continuum. A proper conceptualization of the means-end continuum would be one that delineates terminal program outcomes as means to some acknowledged higher order value implicit in some larger context. This context will almost always be of a higher order or larger magnitude than the program itself. Most often this context comprises the most immediate program, system, or organizational unit of which the program is a component part.

VALUES: THEIR DEFINITION AND FUNCTION
IN PROGRAM EVALUATION

Values represent an important vehicle for getting beyond program objectives in evaluation. The concept of values, however, has traditionally been an amorphous and elusive concept. This has been a result of the complex and multidimensional nature of the concept itself and the relative dearth of effort by evaluators in defining, accepting, and using this concept as a legitimate tool.

Values are factorially complex variables, different from the factorially "pure" variables that evaluators and program personnel typically use. One way in which to conceptualize a value is to think of it as a composite or factor derived from the interrelationships among many needs, opinions, attitudes, and beliefs. No one measure and no one behavior can fully represent a value, just as no one variable in a factor analysis can adequately represent any factor in that analysis. Whether values are arrived at informally or statistically, they are always interpretations of the patterns, consistencies, and regularities that lie beneath behaviors and, hence, constitute abstractions that can be used to represent these behaviors.

The concept of values routinely appears in the literature of sociology, anthropology, political science, economics, the arts, and the humanities, and represents an important area of inquiry within these fields. It has been late, however, in becoming a legitimate focus in the field of evaluation as can be noted from the extensive (and sometimes exclusive) reliance on behavioral objectives, measurement, and the decision-oriented approach during the early development of the field. This is presently less the case, and the time seems ripe for this concept to fulfill its complementary and contributory role in program evaluation. This role should not be surprising as the word *evaluation*, both literally and figuratively, has its roots in the word *value*.

In the broadest of terms, values stand for the things we move toward (e.g., objectives, goals, aims, preferences, aspirations, and expectations) and the things we wish to avoid (e.g., fears, anxieties, apprehensions, and dislikes). Values, especially the collective values of large numbers of individuals, are customarily taken as indications of what a group, subculture, or society believes to be *good, just,* and *appropriate.* In program evaluation values can serve as a link between the program being evaluated and some larger end or value to which the program is supposed to be contributing. Values link means with ends and provide one type of standard against which the worth or merit of program objectives can be tested. Values, then, represent a constituency broader than that of program administrators, evaluators, staff, or program recipients; it includes the entire policy-shaping community, as Cronbach et al. (1980) use the term.

The uses of value data in evaluation can be classified into several general purposes. The first general purpose that values can serve is as standards with which the merit and appropriateness of program objectives can be determined. These standards have their roots in the values that lie beneath the needs, opinions, attitudes, and beliefs of those the program is to directly and indirectly serve. To discover the undergirding values of a program is to discover the rationale upon which standards for judging that program will probably be based. Since standards should not only represent what program planners, administrators, and sponsors may want for a program but also what program participants, the community, and citizens at large may want and expect, standards need to be pluralistic in origin. Since values imply criteria originating outside the narrow confines of the program, they serve as an important means of identifying program standards representing some broader "good" and reflecting a comprehensive view of program merit or worth.

A second general purpose values can serve is to enable a determination of whether program goals are ethical and moral. To determine the ethics of a program, the evaluator might ask if program goals reflect the hopes, expectations, aspirations, needs, and desires of the clients to whom the program is directed? That is, do program goals actually reflect the inner world of program participants as well as what program planners and administrators believe this inner world to be? Vickers (1970) describes this inner world as one fundamentally structured by human values:

> Men themselves are *valuers*. They seek and shun; and their seekings and shunnings are to be understood not in terms of the outer world which the ecologist can observe but in terms of the inner world which his subjects inhabit and which he may or may not share. In any field in which men function, the relevant facts and forces include not only what is happening but also what men think is going to happen; not only what they are doing to each other but what they expect, hope, fear from each other and from themselves. The inner world is fundamentally structured by human values [p. 56].

Programs unresponsive to the inner world of those they purport to serve run the risk of questionable ethics. Sterilization programs, some correctional programs, and many educational programs represent programmatic efforts which often fail to fully appreciate the inner world of program participants, resulting in the unethical treatment of some participants.

To question the morality of a program one might ask if the program being evaluated distributes community and societal resources in ways that are consistent with the values of those whose resources are being expended for purposes of the program; is the amount being spent commensurate with how important the group "footing the bill" sees the problem? What opportunities might be *lost* for program participants, what participants might *not* be served, and what other community or societal interests will *not* be addressed as a result of program sponsors choosing a particular course of action? These are questions concerning the morality of a program that derive from a value base.

A third general purpose that values can serve is to help interpret program goals and their relation to specific program strategies or processes. Many strategies may be a satisfactory means to some goal but only some strategies will be compatible with the values that lie beneath the goal. Values become important adjuncts for deciding which strategies are more suitable than others. Without values as a point of reference, vague and nebulous goals are the principal means of judging the appropriateness of various program strategies and of providing a standard for deciding which strategies are most desirable.

The fourth general purpose that values can serve is to assist the evaluator in determining if program outcomes are consistent with and contributory to the outcomes of some larger program, system, or organizational unit of which the program is a component part. This purpose is intimately tied to the means-end continuum. Do program outcomes represent logical stepping stones to desired outcomes more distant in time and place than the behavior to be measured at program completion? Are program outcomes the best means to these desired ends? Since all programs have a purpose, they are expressions of value. One role of value-oriented evaluation is to explicate these values and to determine the utility of program goals for reaching these more ultimate ends to which the program should be contributing.

NORMATIVE VALUES AND BIAS

For values to be useful in program evaluation they must represent the focused and considered opinions, attitudes, needs, and beliefs of those the program is to serve both directly and indirectly. These opinions, attitudes, needs, and beliefs can be called *normative values*. Normative values are characteristic of large numbers of individuals who have a vested interest in, have thought about, and have independently decided upon a particular course of action. Normative values are similar to the Cronbach et al. (1980) concept of "prevailing views," which they see as the basis for the process of negotiation and accommodation within the policy-shaping community.

Separating normative values from personal biases, however, is not always easy, and yet is absolutely essential to a value-oriented conceptualization of program evaluation. Consider the following, for example:

"Unemployment shall not be allowed to exceed 5% of the labor force." (laid-off autoworker)

"The cost of living index shall not be allowed to rise more than 3% per year." (clerical worker passed over for promotion)

"Adult bookstores shall not be permitted within 300 yards of churches and schools." (clergyman)

"Drug abusers should be given a minimum of 3 years in jail." (state trooper)

"The government should not interfere with an individual's right to bear arms." (hunting enthusiast)

"All students must read at the ninth grade level in order to graduate from high school." (school principal)

Are these expressions of personal bias or are they normative values? Declarations of this kind do provide information about the attitudes, needs, opinions, and beliefs of those who make them. However, they provide little or no clue as to the source, history, or relevance of the values they supposedly represent. They are expressions of value that may or may not: (a) have been invented "on the spot," (b) represent the feelings of significant numbers of individuals, (c) represent the feelings of those for whom a particular program is intended, or (d) represent those of program sponsors, citizens, or society at large whose resources will be expended for purposes of a program. The point is that no single attitude, need, opinion, or belief can adequately represent a normative value, although attitudes, needs, opinions, and beliefs are commonly manifest in expressions of value. Value expressions represent only one small aspect of a normative value (the tip of the proverbial iceberg) and thus, when taken by themselves, may actually constitute erroneous or distorted expressions of the normative values they purport to represent.

Unfortunately, many sources of normative values in a program environment go unnoticed, while many sources of bias are mistaken for normative values. This is sometimes the result of the spontaneous and idiosyncratic nature of the value sources themselves, but often it is the direct result of evaluators being insensitive to or poorly trained in methods of obtaining value data, especially from informal sources. Informal sources of value data represent one of the most underutilized and least appreciated categories of data for obtaining a picture of the normative values operating in a program environment. When taken collectively and when properly scrutinized for bias, these sources can provide sensitive and revealing portraits of the normative values operating in a program environment. The following are some of these sources:

1. Case histories and critical incidents reflecting the needs and preferences of potential program participants

2. Past choices, expressed or actually carried out, of potential program participants related to the services to be provided by a program
3. Opinions reported in the mass media of sponsors, citizens, and participants concerning a program
4. Frequency and types of newspaper and magazine articles related to program content and services
5. Attendance at meetings from which inferences could be drawn regarding the need and potential popularity of a program
6. Frequency and topics of written correspondence received pertaining to a program
7. Social indicators (e.g., vandalism, utilization of facilities, recidivism, etc.) that could indicate need or desirability of a program
8. Participation by potential participants in activities related to the program
9. Frequency of requests for program-related information, materials, and service
10. Telephone calls, praiseworthy or critical, of program-related ideas
11. Budget trends indicating percentage of funds spent on similar programs over time
12. Census data reflecting program-related population changes and characteristics
13. Judgments expressed by constituent groups concerning needed changes in the program environment
14. Attitudes and opinions expressed in interviews of potential program participants
15. Subjective-descriptive consultant or interagency evaluations relevant to the program
16. Reports, anecdotal and otherwise, by visitors assessing conditions relevant to the program
17. Attitude, need, and public opinion surveys from which more direct inferences about values can be drawn

The usefulness of these data to a value-oriented conception of program evaluation depends on the ability of the evaluator to separate normative values from the biases that can be mistaken for them. According to Kaplan (1974), bias exists when one's will to believe is motivated by interest external to the context of the inquiry itself. Ignoring data to the contrary of one's belief is a cardinal feature of bias. Values and biases are not the same. All inquiry is directed by values and is inseparable from them. A "scientific" or "objective" inquiry merely means that the inquiry attempts to control bias. The evaluator cannot eliminate bias, but can attempt to control it, compensate for it, or make it known. The fact that planners, evaluators, policy makers, clients, vested interest groups, and technical experts all have values does not of itself imply that these groups are biased. Values engender bias not when they reveal attitudes, needs, opinions, and beliefs, but when they prejudge solutions and are motivated by concerns external to the purpose, goals, and content of the program.

Several questions are evident in attempting to differentiate personal biases from normative values. The first of these questions focuses on whether the various parties providing the value data have made clear the sources from which their value expressions derive. For example, have vested interest groups and technical experts made explicit the circumstances, objects, or incidents on which their attitudes, opinions, and beliefs are based? If so, are these sources relevant to the program or course of action under consideration or are they external to it?

Another important question that can be asked is whether or not the criteria for interpreting expressions of value are clear—or if there are any criteria at all. Here, the evaluator might determine if the value expressions are presented in such a way that they can be checked against actual program benefits and costs. That is, are the value statements general and diffuse, extending beyond program intents, or are they concrete and specific, relating directly to what the program is trying to do? Few programs can eradicate unemployment, cure drug addicts, raise one's income appreciably, or provide for even a small list of the many legitimate benefits program participants may seek from a program. Such outcomes are generally beyond the scope and focus of any single program and statements espousing such values must be "scaled down" or reinterpreted to match specific program goals before they can comprise useful judgment criteria for program evaluation.

Another question that can be asked is how vested and personal the value data are. Can it be known if personal wishes violate the factuality of things or are the two hopelessly muddled? Can it be known whether a group's unqualified support for a program is the result of a real need or condition or the result of some obvious economic (or other) benefit to those who may wish to be involved with or participate in the program? Can these different motives be determined from value statements alone? If not, what other types of data must be gathered before these statements can become credible? And finally, when these other types of data are gathered do the same value statements emerge? These are not easy questions to answer as each has many different facets and few implicit answers. However, several general procedures suggested by Guba (1978) can make the task somewhat more manageable.

Triangulation. Triangulation involves building a case for value data by determining its consistency with other data sources, research methods, and theoretical schemes. It is the back and forth checking and cross validation of data using various sources to determine if value statements emerge with some consensus. Data from the same sources, even though they may have used different methods, should converge to yield essentially the same conclusions. When incongruencies exist, further value data and other methods should be used to resolve incongruencies. Campbell (1979a) describes this procedure as "multiple methods focused on the diagnosis of the same construct from independent points of observation [p. 63]."

Cross examination. Cross examination involves different individuals in the scrutinizing of the same value data for the purpose of determining whether or not the same conclusions can be derived from them. Do different observers draw different conclusions and inferences from the data? Do their judgments match or do the conclusions change depending on one's frame of reference, perspective, and assumptions brought to the data interpretation task? Incongruencies require identification of the source of the differences. When incongruencies exist other sources of value statements must be sought to clarify the reasons behind the differences.

Convergent validation: Same source, different perspectives. This differs from triangulation in that rather than attempting to use different sources of information, the same source is scrutinized from as many perspectives as possible. These different perspectives might be represented by technical experts from different levels of an agency's hierarchy or testimony derived from individuals with different theoretical persuasions but who would be expected to articulate the same value position. When these different perspectives yield contradictory expressions of value, other data and validation procedures must be employed to resolve the conflicts.

Establishing credibility of the findings. When value statements are informal (e.g., constitute "private" views or discussions with key policymakers or only a small sample of participants) the validity of the data provided must necessarily rely heavily upon the credibility of those from whom the data are derived. This is of course almost always true with the data provided by vested interest groups and technical experts. In such cases credibility checks become important means of examining value data. Credibility checks can be both logical and experiential. For example, the evaluator might inquire into the logic of a value statement by asking, "Does the person providing the value statement have the experience to comment?" or "Has he or she lived and worked in the environment long enough to have developed feelings on the matter and to assess the consequences of current programs and plans?" It is, of course, not unusual for those who speak the loudest to have the least experience or knowledge on which to base their value expressions. Background checks as to prior experience and capabilities of the individuals from whom expressions of value are elicited can be used to cull out those whose contributions might prove marginal or minimal.

Determining credibility experientially means asking individuals already known to be respected and knowledgable to pass judgments on the credibility of the individuals providing value data. Peers can offer the most persuasive evidence as to why particular testimony should or should not be believed. A jury of peers is based upon the belief that peers know best the background, experience, previous biases, and capabilities of those from whom the data are being elicited. Soliciting viewpoints about the credibility of the source from more than two individuals is desirable when the credibility of the peers themselves is unknown. Credibility checks can also be made from previous reports and testimony. Opinions of others about the value of these reports or testimony can be used to judge their consistency and accuracy.

Establishing congruency with referential materials. This approach involves the preparation and use of special materials that can provide information about the values of constituent groups. These materials might include specially prepared videotapes of the actual context in which a new program is to be implemented, or statistical data revealing the performance of the program or similar programs implemented at other sites. Such materials are usually important in that they can be studied repeatedly by various groups and crucial aspects of critical incidents focused on for further inquiry. Degree of congruence is determined by comparing the value statements derived from these referential materials with conclusions derived from any one or any combination of the above methods.

Triangulation, cross-examination, convergent validation, credibility, and congruency are means by which attitudes, beliefs, needs, and opinions can be checked for consistency within and between groups of individuals. Their use is based on the assumption that inconsistencies in value expressions made by either a single individual or group impugns the credibility and, thus, the usefulness of any single value expression. One of the tasks of value-oriented evaluation is to separate consistent expressions of value from those that are inconsistent. It is from this process that a determination of normative values can be made.

This review of procedures for examining normative values should not be interpreted to mean that it is always the task of the evaluator to collect value expressions. These data may already exist in the program environment as the result of the work of planners, needs assessment surveys, goal clarification projects, public opinion polls, and in the goal statements of higher level programs, systems, or organizational units to which the program contributes.

First and foremost, the role of the value-oriented evaluator is to carefully inspect the program environment for such data and to reveal its normative qualities, if present. Failing to arrive at usable value data from these sources, the evaluator may then scan more informal sources with qualitative methods (cf. Patton, 1980) or collect value data with one of the naturalistic methods to be described below. With any of these alternatives, however, the traditional definition of evaluation is considerably broadened to include the act of determining standards or criteria from which a judgment of program merit or worth can be made. Thus, in a value-oriented conceptualization of program evaluation, the evaluator is not only responsible for collecting the data from which judgments of program merit or worth are made but also for explicating the criteria upon which these judgments should be based. As we have seen these criteria derive not from the personal values of the evaluator or from any single source but from the normative values that exist among all parties to the program. These normative values should be, but are not always, consistent with a program's objectives.

It would seem that such an assessment of values would be necessary to conduct a multipartisan evaluation of the type described by Cronbach et al. (1980). If different value perspectives are never discerned, the evaluation has no chance of being truly multipartisan. While granting that the extent to which this value analysis can be carried out is dependent on many factors, it would appear that the evaluator needs to pursue this analysis to the extent that his or her operating context allows. See also Cronbach et al. (1980), Coleman and others (1979), and Mitroff, Emshoff, and Kilmann (1979).

Some Naturalistic Inquiry Models for Collecting Value-Oriented Data

Among the value-oriented methods and concepts implied by Cooley and Lohnes's (1976) updating of Dewey's theory of valuation is that of naturalistic inquiry. An outgrowth of ecological psychology (Barker, 1965, 1968) and the slackening of the taut reins of empirical psychology in the 1960s and 1970s, naturalistic inquiry stands apart from the more formal decision-oriented models of evaluation reviewed in Chapter 1. Naturalistic observation has been referred to as an alternative to conventional evaluation methodology, departing from both traditional forms of instrumentation and traditional methods of data analysis. While naturalistic methods have been used for many different purposes and applied in many different orientations (anthropology, sociology, psychology, and education, to name a few), the perspective and philosophy of naturalistic methods make them ideally suited for systematically observing and recording normative values.

Although several definitions of naturalistic inquiry have been offered, most authors (e.g., Guba and Lincoln, 1981; Patton, 1980) agree that naturalistic inquiry differs from other modes of inquiry by its relative position along two dimensions: (a) the degree to which the investigator manipulates conditions antecedent to the inquiry, and (b) the degree of constraint imposed on the behavior of subjects involved in the inquiry. Accordingly, naturalistic inquiry has been defined as:[1]

> any form of research that aims at discovery and verification through observation [Willems and Raush, 1969, cited in Guba, 1978, p. 4].

> slice-of-life episodes documented through natural language representing as closely as possible how people feel, what they know, how they know it, and what their concerns, beliefs, perceptions, and understandings are [Wolf and Tymitz, 1976-1977, cited by Guba, 1978, p. 3].

> evaluation which attempts to arrive at naturalistic generalizations on the part of the audience; which is aimed at non-technical audiences like teachers or the public at large; which uses ordinary language; which is based on informal everyday reasoning; and which makes extensive use of arguments which attempt to establish the structure of reality [House, 1977, cited in Guba, 1978, p. 3].

In addition, naturalistic studies have been identified by Sechrest as ones which (a) do not require the cooperation of the subject, (b) do not permit the subject's awareness that he is being measured or treated in any special way, and (c) do not change the phenomenon being measured [cited in Guba, 1978, p. 5].

In theory, a naturalistic study consists of a series of observations that are directed alternately at discovery and verification. This process supposedly leads to successive reorientations on the part of the investigator toward the phenomena being observed and to further discovery. Unlike formal evaluation models, the naturalistic inquirer approaches data collection with a minimum of preconceived categories or notions of what will be seen, as though the behavioral phenomena were being observed for the first time. Any effort to manipulate any part of the program setting prior to observation or to constrain the behavior of those being observed reduces the "naturalism" of the method.

How data are tabulated and analyzed in a naturalistic study is left up to the investigator, although it invariably includes some form of observation followed by a piecing together of relationships, patterns, or consistencies in the data. These relationships, patterns, or consistencies are then used to further channel and focus subsequent observations. Data recording methods

---

[1]We are indebted to Guba (1978) and Guba and Lincoln (1981) and the sources cited for some of the material upon which this section is based.

may include impressionistic accounts and ethnographic records[2] of the phenomenon observed as well as more structured observation systems in which data are coded according to some general classifications, although these classifications must be considerably broader than those typically employed by evaluators. From these data still more categories of behavior are derived which are expanded and verified through still further observation.

Naturalistic inquiry has become an important, if not major, component of many present day conceptualizations of program evaluation (Guba and Lincoln, 1981; Murphy, 1980; Parlett and Hamilton, 1976; Patton, 1980; Reichardt and Cook, 1979; Rist, 1980), resulting in the development and description of a sizable array of naturalistic strategies, methods, and techniques. The influence of naturalistic inquiry in this regard has been considerable and represents what might be described as a movement away from the development of conventional evaluation practices and more formalistic definitions of evaluation (for example, the decision-oriented models described in Chapters 1 and 2) and toward a more value-oriented or qualitative conception of program evaluation.

Naturalistic inquiry represents both in spirit and method a value-oriented approach to evaluation. Value-oriented writers such as Dewey, Scriven, and Apple would find solace in the fact that naturalistic inquiry, more than most other methodological perspectives, is likely to yield data unconstrained by preconceived notions about the program and its environment, and what the program is and is not supposed to be doing. This perspective seems congenial to examining means–end relationships (Dewey, 1939), side effects and unanticipated program outcomes (Scriven, 1973), program consequences vis-à-vis external or higher order criteria (Brown, 1970), fundamental issues which question the rational upon which a program is based (Apple, 1974), the various value positions present in the policy-shaping community (Cronbach et al., 1980) and the morality and ethics of a program.

Some evaluators have turned to the naturalistic approach as a result of what are perceived to be serious limitations of decision-oriented evaluation methods, namely that (a) decision-oriented models have been too restrictive in the types of data that can be collected, (b) these models generally have no way of incorporating the concept of values into the decision-making process, (c) decision-oriented evaluation may at times actually contrive data by defining or manipulating dimensions of a program which have no practical meaning in the real world, and (d) decision-oriented models may actually

---

[2]A type of observational record associated with the field of anthropology in which behavior is recorded in relation to the context in which it occurs and is ascribed meaning only in relation to this context.

constrain and, therefore, distort through formal instrumentation the responses expected of subjects.

In response to these limitations, several evaluators have developed "alternative models" or approaches to evaluation which embody the elements of naturalistic inquiry. We offer these models (cited in Guba, 1978) as examples which, to varying degrees, embody the elements of value-oriented inquiry. These models or approaches allow for the observation, description and verification of the standards, criteria or values operating in the program environment which may illuminate decisions that need to be made about a program.

THE RESPONSIVE MODEL

The first model with some relationship to value-oriented inquiry is the responsive model developed by Stake (1975a,b). The responsive model focuses on important issues and concerns pertaining to a program. According to Stake, evaluation is responsive if it "orients more directly to program activities than to program intents, responds to audience requirements for information, and if the different value perspectives are referred to in reporting the success and failure of the program [1975b, p. 14]."

The primary purpose of responsive evaluation is to respond to audience requirements for information and to bring to the foreground different value perspectives that might be held by different audiences. Its methodology, like naturalistic inquiry itself, is nonconstraining. Stake describes it in the following terms:

> To do a responsive evaluation, the evaluator conceives of a plan of observations and negotiations. He arranges for various persons to observe the program and with their help prepares brief narratives, portrayals, product displays, graphs, etc. He finds out what is of value to the audiences and gathers expressions of worth from various individuals whose points of view differ. Of course, he checks the quality of his records: he gets program personnel to react to the accuracy of the portrayals; and audience members to react to the relevance of his findings. He does much of this informally—iterating and keeping a record of action and reaction. He chooses media accessible to his audiences to increase the likelihood and fidelity of communication. He might prepare a final written report, he might not—depending on what he and his clients have agreed on [cited in Guba, 1978, pp. 34-35].

These activities are carried out in a series of steps which may be described as (a) talking with clients, program staff, and audiences; (b) identifying program scope; (c) providing an overview of program activities; (d) discovering purposes and concerns; (e) conceptualizing issues and problems; (f) identifying data needs relevant to the issues; (g) selecting observers and instruments (if any); (h) observing designated antecedents, transactions, and outcomes; (i) thematizing or preparing portrayals in case studies; (j) winnowing, matching issues to audiences; (k) selecting a format for audience use; and (l) assembling formal reports (if any).

THE JUDICIAL MODEL

A second naturalistic model with some relationship to value-oriented inquiry is the judicial model. Developed by Wolf (1975, 1979), Owens (1973), and Levine (1974), the judicial model is patterned after the administrative hearing in a court of law. The purpose of the judicial model is to inform and adjudicate issues related to the object or activity being evaluated. Advocates or counsels take opposite views with respect to an issue and argue as convincingly as possible their side of the issue. Judge and jury hear testimony from "witnesses" and the presentation of facts regarding the issue, then offer their opinions as to the merit or worth of the program and their recommendations for improvement. Like the judicial process itself, this approach to evaluation assumes that "truth" (and values) are more likely to emerge in an adversary setting than in the case of a single evaluator using conventional evaluation models and data collection methods.

Generally, the following steps are employed in the judicial model:

1. Issue generation. The issues are identified through "fact-finding interviews" with samples of the audiences involved, as in the case of the Stake responsive model.

2. Issue selection. The purpose of this stage is to delimit the number of issues and to prioritize them, so that they will be manageable in a hearing format.

3. Preparation of formal arguments. Each counsel or advocate team prepares formal arguments related to the selected issues. Available evaluation or other data may be used (to be introduced as "exhibits" in the hearing stage), and additional evidence may be collected, particularly evidence in the form of depositions from witnesses. Additionally, selected witnesses may be asked to give testimony at the hearing itself.

4. Pre-hearing discovery sessions. Each advocate team reviews the major arguments it intends to make and discloses the main features of its "evidence" for the other. Since the hearing is not a "trial" in the conventional sense, but an effort to determine "truth" as precisely as possible, each side shares its findings with the other so that the hearing may be as comprehensive as possible. In addition, the advocate teams decide on ground rules (e.g., number of witnesses to be called and criteria for determining admissability of evidence).

5. The hearing. Modeled on an actual courtroom process, the hearing involves an administrative officer and a "jury" or hearing panel. After hearing the evidence, the jury carries out whatever tasks the advocate teams previously agreed to assign to it, which usually involves at least the determination of findings (which may include judgments of worth) and the making of selected recommendations [Guba, 1978, p. 36–37].

THE TRANSACTIONAL MODEL

A third naturalistic model with some relationship to value-oriented inquiry is the transactional model described by Rippey (1973). This model differs from conventional models in that it deals directly with management conflicts and institutional change brought about by the implementation of a program, utilizing what its authors call "open systems theory." Transactional evaluation studies institutional disruptions brought about by the program and works to ameliorate these disruptions through strategies for conflict management.

Transactional evaluation has five phases:

1. The initial phase. Pre-existing unrest or some other troublesome situation exists. A meeting is set up of interested parties under the direction of a "neutral" evaluator working in a nonjudgmental atmosphere.
2. Instrumentation phase. During this phase, a *Transactional Evaluation Instrument* (TEI) is developed whose purpose is to provide the evaluator with insight into the perceptions and expectations of various interest groups. The instrument also provides a forum for the sharing of opinions among the groups. The TEI is developed and administered in group sessions, during which (a) the evaluator initially formulates issues on the basis of general expressions from the group, (b) participants are asked to re-express opinions about them, (c) the most representative and divergent of the written responses are carefully worded into items that can be rated on a scale from "strongly agree" to "strongly disagree," (d) the instrument is administered to the group, and (e) responses are examined.
3. Program development. The program is redefined to reflect those goals and values on which the group can achieve some consensus.
4. Program monitoring. Various groups agree to assume responsibility for implementing and monitoring the developed program.
5. Recycling. As new conflicts emerge, the entire process is recycled to whatever phase is appropriate [Guba, 1978, p. 38; Talmadge, 1975].

THE CONNOISSEURSHIP MODEL

A fourth model with some relationship to value-oriented inquiry is the connoisseurship model developed by Eisner (1979, 1975). This approach views educational evaluation as a form of criticism. In Eisner's view, criticism depends upon connoisseurship—or the private act of appreciating and sensing the subtle qualities of an object or activity. *Critical guideposts* used to conduct the evaluation are essential elements of the connoisseurship approach. These guideposts represent the values and concepts formed from tradition, experience, and theories about the standards for judging the object or activity. Guba (1978) characterizes connoisseurs as:

> persons with refined perceptual apparatus, knowledge of what to look for, and a backlog of previous relevant experience. They have the ability to recognize skills, form, and imagination and to perceive the intentions and leading conceptions underlying the entity being evaluated. In effect, because of these characteristics, the connoisseur is himself the evaluation *instrument*. Having made his judgments, he communicates the qualities that constitute the entity being evaluated, its significance and the quality of experience engendered by interaction with it, often through the use of rich metaphors [p. 39].

THE ILLUMINATION MODEL

Another model representative of value-oriented inquiry is the naturalistic model developed by Parlett and Hamilton (1976). This approach to evaluation relies heavily on open-ended observations (but also questionnaires, interviews, and tests) to continuously record ongoing events in order to identify (a) critical and nonobvious characteristics of a program, (b) the tacit assumptions underlying it, (c) interpersonal relationships affecting it, and (d) complex realities surrounding the program. In the authors' words:

> *Illuminative evaluation* takes account of the wider contexts in which programs function. Its primary concern is with description and interpretation rather than measurement and prediction. It stands unambiguously within the alternative methodological paradigm. The aims of illuminative evaluation are to study the innovatory program; how it operates; how it is influenced by the various school situations in which it is applied; what those directly concerned regard as its advantages and disadvantages; and how students' intellectual tasks and academic experience are most affected. It aims to discover and document what it is like to be participating in the scheme, whether as teacher or pupil, and in addition, to discern and discuss the innovation's most significant features, recurrent concomitants, and critical processes. In short, it seeks to address and to illuminate a complex array of questions [Guba, 1978, p. 40].

Illuminative evaluation is carried out in three stages:

1. Initial observations for the purpose of familiarization with day-to-day reality of the setting(s), largely in the manner of social anthropologists or natural historians
2. More sustained and intensive inquiry into a number of common incidents, recurring trends, and issues frequently raised in discussion

3. Efforts to seek general principles underlying the organization of the program, determine patterns of cause and effect within its operation, and place individual findings within a broader explanatory context [Guba, 1978, p. 40].

NATURALISTIC VALUE-ORIENTED MODELS—A SUMMARY

The models described qualify as naturalistic insofar as they adhere to the two primary conditions set forth by proponents of the naturalistic method: (a) they do not manipulate conditions antecedent to the inquiry, and (b) they pose minimal constraints on the behavior of participants involved in the inquiry. Although always a matter of degree, these five models meet these conditions to a greater extent than do most conventional approaches to evaluation.

The most significant aspect of naturalistic inquiry in program evaluation, however, may be its capacity to uncover values that can constitute the criteria for determining program merit or worth. These values can indicate what data to collect, how best to collect them, and how they are to be interpreted. In explaining the importance of values in making these decisions, Kaplan (1964) compares a value-free inquiry (or one that limits itself just to describing what objectively happens) as the "dogma of immaculate perception" and compares it to the position of the aesthetes at the turn of the century who viewed art as a matter of pure form or decoration, "at the cost of making it an idle song for an idle hour [p. 385]."

Finally, it is important to note the perspective or mind set the value-oriented evaluator carries with him or her when studying values. This perspective, or *Weltanschauung,* has been aptly captured by Louch (1966), who provides a good portrayal of the value-oriented inquirer and the commonality between the naturalistic and value-oriented approaches to evaluation. He views the process of uncovering values as one in which:

> behavior cannot be explained by a methodology borrowed from the physical science . . . . what is needed . . . . is not measurement, experiment, prediction, formal argument but appraisal, detailed description, reflection and rhetoric. . . . Human action is a matter of appraising the rightness or appropriateness of what is attempted or achieved by men in each set of circumstances. Its affinities are with morality rather than with the causal or statistical accounts appropriate to the space-time framework of the physical sciences. Its methods are akin to the deliberations and judgments in the law rather than the hypotheses and experiments of physics [cited in van Gigch, 1978, p. 220].

In essence, the value-oriented approach holds that the actions of people and the programs they create must ultimately be justified in terms of values. This perspective as well as the application of naturalistic and other qualitative methods to program evaluation, however, does not preclude the use of measurement, experiment, prediction, and formal argument—the so-called quantitative approach. Each of these forms of inquiry used at the proper time and for the proper purpose can inform and complement the other.

Values: A Broader Context

Many general methods of data collection such as surveys and questionnaires, Likert scales, semantic differentials, content analysis, Guttman scales, Q-sort methods, interviews, and naturalistic records can be applied within a value orientation. However, one general technique that is *directly* grounded in a value-oriented philosophy is that of needs assessment.

THE RELATIONSHIP OF NEEDS ASSESSMENT
TO VALUE-ORIENTED INQUIRY

Kaufman (1972, 1977) gives top priority to the needs assessment approach to evaluation and suggests that needs assessment is an activity that should be conducted at all stages of program planning and evaluation. In Kaufman's writing (Kaufman, 1977) six types of needs studies are posited. The functions of these six types of needs studies are to (a) identify problems based upon needs (Alpha type), (b) determine solution requirements and identify solution alternatives (Beta type), (c) select solution strategies from among alternatives (Gamma type), (d) implement program (Delta type), (e) determine performance effectiveness (Epsilon type), and (f) revise as required (Zeta type).

To Kaufman evaluation is a sequential series (Alpha through Zeta) of needs assessments. Because of their relevance to the following discussion, the characteristics and assumptions pertaining to each of these six levels are shown in Table 3.1.

Kaufman's Alpha needs assessment is an analysis of the difference or discrepancy between "what exists" and "what should be." An Alpha needs assessment precedes any consideration of a program and is intended to determine the goals, objectives, and policies upon which a program can be based. For Kaufman an Alpha needs assessment is a "no-holds-barred" strategy that pays no deference to the status quo. The results of this type of needs assessment may well point to the abolition of some existing programs or the development of new programs in areas that do not readily fit with the existing political, economic, or social structure.

In Beta needs assessment, program goals and objectives are "givens" (having been specified in an Alpha needs assessment) and emphasis is

Table 3.1
Kaufman's Taxonomy of Needs Assessment[a]

System approach functions	Needs assessment type	Possible characteristics	Possible assumption base
Identify problem based upon needs	Alpha	External criteria, partnership-based, utility referent for survival and contribution; single emphasis upon "need" as an outcome gap.	Almost anything may be changed and questioned; there can be no sacred cows; even laws can be added, deleted, modified.
Determine solution requirements and identify solution alternatives	Beta	Partnership-based, analysis of performance gaps within the system, analysis of process and solution gaps; focus on outcome as well as process gaps.	Work is to be conducted within a context, usually organizational; and, for the most part, the rules and policies of the organization, as they now exist, are the ground rules for planned change.
Select solution strategies from among alternatives	Gamma	Ranking of solutions by partners. Cost-efficiency models, cost-effectiveness models, etc.	The existing goals and objectives are useful and appropriate, and the charge is to find the most effective manner to meet the objectives.
Implement	Delta	Determination of gaps in prespecified performances; management-by-objectives, management-by-exception, scheduling, etc.	It is known what is to be done and how to do it, the important function here is to successfully administer the jobs to be done and the resources to help get the jobs accomplished.

(*Continued*)

placed on the processes to be employed in reaching these goals and objectives. Here the general direction for any new program or program modification is already established; yet to be decided are the most effective and practical strategies by which established goals, objectives, and policies can be accomplished.

For a Gamma needs assessment, Kaufman's third category, both goals and objectives and the programs to be utilized in reaching these goals and objectives are givens. What still must be accomplished, however, is the matching of different goals with different programs and the prioritization of

Table 3.1 (*Continued*)

System approach functions	Needs assessment type	Possible characteristics	Possible assumption base
Determine performance effectiveness	Epsilon	Determine discrepancies between results and objectives for the end-of-term/project/program for decisioning.	The jobs have been done, this function is to determine the gaps between the goals and objectives and the accomplishments.
Revise as required	Zeta	En route evaluation of both processes and progress toward outcomes.	While the jobs are getting done, or after we have finished any segment of one or more jobs, discrepancies between our goals and objectives (usually en route) are determined, and corrective action is instituted, or a decision not to change is accepted and implemented.

[a] Used with permission of Educational Technology Publications.

goals and programs according to their perceived importance or need. This type of needs assessment requires little developmental activity as the focus is on matching up goals and programs that already exist and identifying the most valued from among them.

The remaining three levels of needs assessment described by Kaufman depart considerably from the nature and purpose of the first three, and one might even question if the latter should be considered types of needs assessment. This question leads to an examination of Kaufman's rationale for including in his taxonomy the Delta type (program management), Epsilon type (summative evaluation), and Zeta type (formative evaluation) needs assessments. With some thought it becomes obvious that the rationale underlying Kaufman's entire taxonomy is not one of describing types of needs assessment but rather the process of planning and evaluating. Viewing Kaufman's taxonomic structure as strategies for planning and evaluating, the Alpha, Beta, and Gamma type needs assessments emerge as strategies for planning a program and the Delta, Epsilon, and Zeta type needs assessments emerge as strategies for evaluating a program. All six levels of Kaufman's taxonomy can be described as methods for bringing about planned change. Kaufman's taxonomy simply represents an orderly and successive array of methods for accomplishing change.

A further point concerning Kaufman's taxonomy is that it contains an internal–external dimension. Beta through Zeta needs assessments can be considered internal strategies for planned change in that the goals, objectives, and policies on which a program is to be based (Beta) and even the program itself (Gamma–Zeta) are already determined. If change is to result from any one or a combination of these strategies it must be a product of certain "givens" which are internal to the structure if not the content of the program to be planned. An Alpha needs assessment, however, is different. This was referred to earlier as a "no-holds-barred" strategy for which Kaufman states there can be "no sacred cows." That is, existing goals and objectives might well be ignored as a result of discovering new goals and values to which the status quo is unresponsive. An Alpha needs assessment can be conceptualized as the activity with which all planning begins. This beginning point is an assessment that determines the goals, objectives, and policies for planning a program and, hence, is independent of any program. An Alpha needs assessment is, in essence, an assessment of the values upon which the program should be based.

Kaufman's taxonomy of needs assessments brings us full circle in our value-oriented discussion of program evaluation. Program evaluators generally begin with the assumption that the program is a means to some end. That is, they accept program objectives as givens and then proceed to determine if the program is an effective means to the end stated in those objectives. However, Kaufman's Alpha needs assessment leads to the question of why the program was developed in the first place. Does the end derived from program objectives relate back to the goal, ameliorate the need, and is it congruent with the *values* that provided the basis for the program in the first place? These values are the most logical beginning point for analyzing the means–end continuum.

A beginning point external to the program is critical if the evaluator is to avoid confusing means with ends. To the value-oriented evaluator, outcomes relevant to these objectives are only means to still further ends and it is these higher order ends that simultaneously provide the criteria from which program objectives must be derived and program outcomes evaluated. It is by applying these higher order ends to the evaluation of competing program strategies and alternatives that relative judgments of worth can be made.

This perspective also coincides with Kaufman's (1977) view with respect to making faulty assumptions in achieving planned change. Kaufman states that the evaluator could start at any level of needs assessment but the further from the top (Alpha type) the level of entry, the lower the probability of actually achieving a meaningful change in practice, and the greater the probability of making errors due to faulty assumptions. Value-oriented definitions of evaluation strongly suggest entry at the Alpha level (to identify values and design the program). Decision-oriented definitions generally assume a lower entry level, usually Epsilon and Zeta type needs assessments (to develop, manage, and evaluate the program).

Before leaving the topic of needs assessment, a caveat is in order. We have emphasized Kaufman's work in discussing the relationship between values and needs assessment primarily because his Alpha needs assessment contains an implicit theoretical rationale for the activities contained in value-oriented evaluation. However, we do not completely agree with Kaufman's definition of a need. For us *true* needs can only be derived from values, not simply from the perceptions that individuals have of "what exists" and "what should be." Wants, ideals and perceptions of what should be can not be taken as the sole basis of needs, as these are often impractical, sometimes even unattainable. Although felt desires, wants, and even ideals can be legitimate signs of true needs, it is the higher order and multifaceted concept of normative values that is indispensable in determining need. Normative values are more ubiquitous and less susceptible to direct measurement than needs, often encompassing both direct and indirect data from many different sources (see pp. 64-65, this chapter). For needs assessment work in the human service field see Bell, Warheit, and Schwab (1977); and Siegel, Attkisson, and Carson (1978).

VALUES, PLANNED CHANGE, AND INCREMENTALISM

As just noted in our discussion, one role of planning and evaluation is to bring about planned change. There are, however, a number of contrasting views as to the best way of accomplishing planned change. Two of these views are particularly germane to our discussion, in that one embodies a value-oriented perspective while the other is the antithesis of this perspective, which we will refer to broadly as incrementalism.

The incrementalist view of planning and evaluation has been most cogently put forth in a book by Braybrooke and Lindblom (1963). These authors articulate a strategy for planning and evaluating in which improvements are made in a system or program setting by measuring increments of output that may be produced by different alternative courses of action. The developer and evaluator proceed not by defining the "best of all worlds" but by calculating the types of modifications that can be made *within the existing system* that would result in an improvement over the status quo, taking into consideration as much as possible that an improvement in one sector of the system could result in an actual decrement in performance in another sector. Incrementalism is a remedial process whereby decision makers essentially deal with shortcomings in the program environment by determining how much of one desired outcome must be sacrificed in order to attain an increment of another desired outcome. In essence, this position holds that planned change is best achieved by choosing a particular course of action from among alternatives that differ only in small degrees from the status quo.

To illustrate this point let us consider an example in which a developer and evaluator have been requested to help revise an existing program designed to teach its participants both skills and concepts, the former representing behaviors related to on-the-job performance and the latter representing understandings, conceptualizations, and judgments related to these behaviors. Moreover, let us assume that some direct pressure has been applied for the program to teach some additional skills. One way in which to proceed would be for developer and evaluator to determine if the additional skills could be taught using the existing resources and format of the program. That is, they might first ascertain if there is sufficient "slack" in the program whereby new skill objectives could be added without incurring some cost (i.e., without creating a decrement in some other objective). It is of course unlikely that no loss of any kind would be incurred, particularly if resources and the general format of the program were to remain reasonably fixed.

It is at this point that differences in the *incremental* and *value-oriented approaches* would first appear. The incrementalist begins with the notion that the program must remain essentially intact and that the best way to proceed would be to make one or more incremental modifications to the program that could accommodate the increased emphasis on skill development. The incrementalist determines the *gains* and *losses* that might be incurred by various program modifications that would incorporate this new emphasis. Once the decision is made to keep the "whole" essentially intact, the incrementalist generally must count on some losses, and so some effort must be devoted to establishing the *marginal* increases (gains after losses have been subtracted) that would result from pursuing a number of different modifications. In our example this might result in an attempt to determine the marginal gain that might be realized if one or more new skills but one less concept were taught.

And it is here the incrementalist confronts the issue of values in the program setting. Values are dealt with, however, not as vehicles for elucidating the means-end continuum (i.e., explicitly exploring the values operating in some larger context to which the program is supposed to be contributing) as would be the case with a value-oriented perspective. Instead, values are viewed as vehicles for determining "marginal" increases that might be achieved by using strategies close at hand and that require the least reorganization of the whole. In this manner, the incrementalist eliminates the least desirable alternative but may not arrive at the most desirable one. For this latter outcome to occur, the developer and evaluator would have to have a conception of alternative strategies linked not to themselves but to some higher order end that is consistent with the values that lie within some larger context. In short, the evaluator would have to have a conceptualization of the means–end continuum for this particular program setting. The philosophy of program planning and evaluation we have described has been defined as a strategy in which the evaluator

> concentrates his evaluation on what we call margins or increments, that is, on the increments by which value outputs or value consequences differ from one policy to another. He need not ask himself if liberty is precious and, if so, whether it is more precious than security; he need only consider whether an increment of one value is desirable and whether, when he must choose between the two, an increment of one is worth an increment of another. In essence, each strategy is to produce more of something believed to be desirable, thus the strategy is margin-dependent, i.e., dependent on marginal increases [Braybrooke and Lindblom, 1963, p. 85].

To the value-oriented evaluator, the incremental approach is tantamount to addressing the most felt pain first, whether or not it results in any improvement in the long run. The incrementalist need not subscribe to any larger means–end continuum but rather devotes effort to choosing among means vis-à-vis some immediate goal which is valued but perhaps only in the short run. Even when needed program modifications ostensibly create undesirable deficits in other sectors of the program, the larger means–end continuum still may not be considered. If the evaluator's action on subproblem X creates a problem in area Y, this new problem can be attended to by next turning to area Y. If what is done in area Y creates a problem in area Z, it can, in turn, be remedied. Thus, program modifications are confronted through a sequence of remedial attacks on specific problems. As the authors of this strategy state, "the problem of coordination between the program and any larger context to which it contributes need not be attacked explicitly [p. 131]." Understandably this approach to planning and evaluation has been coined *disjointed* incrementalism.

The popularity of this strategy grows out of the appeal of utilitarianism, immediacy, and expediency. These are the same "culprits" which contributed to the growth of decision-oriented evaluation. There is little doubt that this is and will continue to be an appealing strategy, as the context for decision making and the arena of values are always close at hand. Strategies for program planning and revision are chosen from the most immediately obtainable. Such strategies are desirable from the incrementalist point of view because they change the whole little and preclude any need to understand the manner in which the program relates to any larger program, system, or organizational unit of which it is a part. When the issue of coordination and the relationship of the program to some criteria external to the program itself is ignored, any focus on values becomes a focus on "reme-dial" values. Remedial values represent short-term, immediate outcomes that may only partially represent more comprehensive, longer-term out-comes actually desired in some larger context in which the program must operate and to which it is expected to contribute. Remedial values limit attention to ends that are "in view," and these most often are those provided by program objectives. The incremental philosophy can be appealing to the planner and evaluator in that the context of decision utilizes the status quo and avoids the messy issue of values in any larger context. A conception of a means–end continuum and efforts to attribute merit or worth to a program based in part on its contribution to some higher-level goal or value is not a perspective shared by the incremental philosophy.[3]

Table 3.2 summarizes seven ways of characterizing the incrementalist philosophy. Within these characteristics lie the seeds of an approach to program evaluation that is opposite to a value orientation.

SOME EXAMPLES OF HIGHER-ORDER CRITERIA

Brown (1970) has provided some suggestions for several types of criteria that extend the means–end continuum beyond even those that might be discovered with an Alpha needs assessment. Brown's criteria go beyond the goals of the higher-level program, system, or organizational unit of which the program is a part to include more general societal values to which a program should be responsive. Although it may not always be possible to obtain consensus on these criteria, they serve to delineate the outermost boundary of the means–end continuum that is practical for determining a program's worth or merit using criteria external to the objectives of the program. Brown has suggested five separate categories of criteria which may be interpreted as an attempt at explicating broad, societal standards for both planning and evaluating programs. The following are his classifications and some extensions we have made of them to program evaluation.

(1) Whole person growth, which develops the individual's value system

This classification suggests that social, educational, and human service programs should play a role in reinforcing, extending, and elaborating indi-vidual and societal values. Under this classification programs might be ex-

---

[3]This necessarily has been a cursory treatment of the disjointed incrementalist concept. For a more complete account see Braybrooke and Lindblom (1963), Lindblom (1965), Lindblom and Cohen (1979), and Cronbach et al. (1980). See also van Gigch (1978, p. 572) and Church-man (1968, p. 215ff) for a critique of it.

Table 3.2
Characteristics of Disjointed Incrementalism

Characteristic	Definition
Margin-dependent choice	Comparing alternatives all of which are similar to the status quo
Restricted variety of policy alternatives considered	Attending to a smaller variety than all of the possible options that might be imagined
Restricted number of consequences considered for any given option	An explicit restriction of the consequences of any single option because to consider other consequences would be to go beyond the sector of interest
Adjustment of objectives to policies	Adjusting ends to accommodate available means
Reconstructive treatment of data	Redefining problems in the course of solving them so that they fit preconceived boundaries and methodologies
Serial analysis and evaluation	Problem solution consists of a series of steps in which analysis precedes evaluation for a given portion of the problem and then precedes evaluation of a second portion of the problem and so on
Remedial orientation of analysis and evaluation	Approaching problems from the perspective of how to reduce most felt "pain" first as opposed to defining what the ultimate goal should be and how to design the context in such a way as to prevent the problem from recurring

pected not only to produce in participants the knowledge, behaviors, or understandings stated in their objectives but to accomplish this in ways that communicate and protect both societal and cultural values. This might mean not only that a program would be expected to be consistent with the values of its individual participants but that some program activity should be devoted to linking these values to the larger culture of which the program participants are a part. Program goals and strategies that contradict individual, subcultural, or societal values might count against the program even if what is happening to the program participants is exactly what the program objectives state should be happening to them.

(2) Specialized person-growth, concern with the development of motivation and skill, and in-depth understanding of specific areas of knowledge and expertise

Unfortunately, this type of outcome is sometimes the only point of entry for many program evaluations. These evaluations ask "on the average, did the program produce the changes specified by its objectives?" A less direct and seldom studied implication of this type of outcome is the range or variability of behavior produced by the program. Should it count against the program if some individuals, and maybe only one individual, actually "lose ground" as a result of the program, even when the effect of the program on most participants is overwhelmingly favorable? This would be a case in which the mean or average used in the service of program evaluation might mask the fact that the program is unintentionally "victimizing" some program participants in order to benefit others. Under this classification, data on the variability of responses would be made public and taken into account in any final evaluation of the program. This classification also suggests that while the program may legitimately have quite different effects across individuals, if a participant does not gain *any* of the intended knowledge, understandings, or behaviors as a result of the program, the program must provide other benefits for these individuals in order to justify their time and effort with the program.

(3) Growth in the pool of knowledge, which is the discovery of new relationships and new phenomena

This classification suggests that all programs should produce some kind of generalizable knowledge, regardless of how context specific and narrow of purpose their objectives may be. This criterion suggests that generalizable knowledge in program evaluation may not be limited as much by the nature of a program's objectives, as by the use of evaluation practices that fail to consider larger policy issues that could be illuminated by an evaluation of the program. Such issues could be articulated and discussed prior to the evaluation to identify possible areas where broader policy issues and program objectives overlap. This is the approach to evaluation advocated by Cronbach *et al.* (1980) wherein the program is seen as an implementation of policy and the evaluation as an attempt to elucidate the social problem to which the program is responding. As Cronbach and his associates state, the audience for the evaluation is the whole of the policy-shaping community. This community includes policymakers, the project director, operational personnel, illuminators, administrators, and constituents who are likely to profit more from ideas about the social problem under discussion than from "exact data about the program who's star is past its zenith [p. 368]."

(4) Growth in society at large, which is concerned with the application of knowledge to societal problems

Here a value-oriented conception of program evaluation might suggest that the theme or purpose of a program address some problem that is seen as a problem by more than program participants, sponsors, planners, and evaluators. That is, the program should attempt to address some need that on its face relates to symptoms, deficiencies, and ills that are discernable in the society at large and that the reason for the program's existence be, in part, related to redressing these larger-order symptoms, deficiencies, and ills. Programs may sometimes violate the spirit of this criterion in subtle ways. For example, if eligibility requirements for a program are so restrictive that they disqualify some members of the larger society in need of the program, this should count against the program. Such might be the case if, for example, a program purporting to serve all mentally retarded children limited its clientele to those initially responsive to treatment; or if a drug abuse program purporting to serve all offenders took only the least hostile of clients. Consider the present food stamp program. While intended to help ensure the nutrition and health of societal members meeting certain economic criteria, participation in the program requires transportation and cash money. Food stamps are "sold" to recipients, and only at selected sites (usually post offices). A truly destitute person is not likely to have either transportation or a cash reserve, thus limiting participation in the program. As an extension to this classification one might even argue that those in the larger "society" (e.g., citizens, teachers, parents, taxpayers, etc.) who are indirectly served by the program should logically be expected to be better off as a result of the program, not just those individuals who happen to have participated in it.

(5) Growth from involvement in the institution and its environment. This growth can be associated with the positive results of being involved in a program. There are satisfactions, joys and benefits, "psychic and real," resulting from this involvement.

This classification reflects a belief that there should be a balance between the benefits a program provides its participants and the restrictions it must necessarily place upon them in the process of providing these benefits. Even though some programs may not be compulsory or coercive, parts of them can be, if, for example, a program is voluntary but highly restrictive, limiting, and manipulative once the choice to enter the program is made by the participant. The United States Army and heroin detoxification centers stand out as examples. It is the manner in which participants are introduced to and treated in these parts of the program that might count for or against the program.

Even when a program is voluntary there can be those who may be unintentionally harmed directly or indirectly. The manner in which this occurs can be subtle and might include taking away time the participants would have spent more fruitfully on other activities, diminishing their disposable income as a result of paying for a service that does not benefit them, producing negative feelings because the fruits of the program did not meet expecta-

Value-oriented criteria	- 1					Mid-scale				+ 1

Does the program reinforce, extend, or elaborate:
    Agency or institutional values?
    Community or regional values?
    Society or cultural values?										

Does the program assist in relating the values of individual program participants to broader community or societal values?										
Do any program strategies seem to run counter to or contradict accepted community or societal values?										
Does the program have widely disparate effects across individuals? That is, does it engender a large range of or variability in participant behavior, leaving some participants without any tangible benefits at all?										
Did any individual or group of individuals actually "lose ground" as a result of the program (e.g., were any posttest scores lower than pretest scores)?										
Are there any positive "side benefits" to the program that can compensate individuals who fail to benefit from the program in traditional ways?										
Does the program produce any generalizable knowledge that can enhance our understandings and contribute to our knowledge base in areas other than those directly affected by the program?										
Does the program test, explore, or otherwise illuminate any outstanding social or behavioral issues? That is, does the program address some need that relates to symptoms, deficiencies, and ills that are discernable in the larger society?										

Figure 3.1. A framework for considering value-oriented criteria.

Value-oriented criteria	-1				Mid-scale				+1
Is the program restrictive, limiting, or manipulative of participant responses and expression?									
Do program strategies allow participant involvement in any of the following ways: Sharing program goals and objectives? Eliciting opinions about the effectiveness of the program or how future implementations might be improved? Allowing participants to choose from a mix of program activities based upon individual needs and circumstances? Fostering comraderie and social cohesion?									
Does the program distribute fairly and equitably community and societal resources in proportions consistent with the values of the larger "society" whose resources are being expended for purposes of the program?									
Does the program seem to preclude other important opportunities, client groups, and goals as a result of choosing this particular course of action?									
Are program objectives supportive of or contributory to the objectives of some larger program or system of which it is a part? That is, do program outcomes constitute logical stepping stones to desired outcomes more distant in time and place than the behavior being measured at program completion?									

Figure 3.1. (*Continued*)

tions, etc. These program effects can often be measured as *opportunity costs* (i.e., the cost of not being able to do something else as result of the time, money, or physical and emotional energy spent on the program). This classification reflects the belief that it is often the strategies and procedures more than the goals and purposes of a program that affect the happiness, joy, or satisfaction one experiences from a program. For those who do not benefit in more traditional and expected ways from the program, this classification could be one of the most important justifications for the program.

One aspect of this criterion that is seldom considered in program evaluation is the participatory nature of the program. Participants may receive "joy, happiness, or satisfaction," in part, as a result of their actual participation in the program. At one level a feeling of participation may simply be engendered by sharing program goals and purposes with participants, at another level by eliciting their opinions as to how the program might be improved, at still another by actually structuring the program in such a way as to allow participants to choose their own mix of program activities based upon their own needs and circumstances, and at still a higher level by the sense of comraderie and social cohesion fostered by the program. In other words, the level of participatory involvement that can be experienced from a program and the "psychic and real" outcomes that may result from it should be expected to justify using the word "participant" when referring to those individuals who take part in a program.

The basic idea behind Brown's value-oriented criteria is that some standard external to the program itself (i.e., external to the stated needs, objectives, and policies with which the program was designed) should be employed to test the appropriateness of the program. This was the same notion behind Kaufman's Alpha needs assessment. These external criteria need not be based on grand abstractions (cf. Hoban, 1977) but can be derived from the values, goals, and aims of the participants, institution, and community for whom the program is planned. Some of the more salient questions for program evaluation that derive from these value-oriented criteria are summarized in Figure 3.1.

Summary

We have implied in this chapter that there are at least two meanings of the word *value*. One of these meanings applies to the merit or worth of something that is manifest when we say that something is valuable. A second meaning applies to the criteria or standards upon which this judgment of merit or worth is based. This second usage does not imply that we personally value the object or that it even has value in some objective sense but that we wish to know *the criteria* with which the object will be judged. That is, we wish to know the extent to which it is a just, proper, and appropriate means to some desired end. It is to this second usage as well as to the first that the value-oriented evaluator is committed, committed not only to determining the discrepancy, if any, between a program and the criteria established for it but also to identifying the criteria by which the program is to be judged. These criteria lie in the normative values of all those who have a stake in the program. Only in the very best of circumstances will these normative values be clearly discernable. Of course, they may or may not be represented by the program's goals and objectives.

A value-oriented definition of evaluation begins with the premise that program objectives seldom reflect all the criteria with which judgments of program merit or worth should be made. A value-oriented conception of program evaluation presupposes that when this broader set of criteria are ambiguous, contradictory, or unknown, they need to be objectively determined. We gain from a value orientation the notion that program evaluation is an activity that includes not only the collection of data from which judgments of program merit can be rendered but also the process of explicating the criteria upon which these judgments are based.

This theoretical perspective has practical implications for the ways evaluation should be conducted which are in some ways at variance with traditional approaches. Inherent in this perspective is a call for new methods and new concepts in the field of program evaluation leading to an expanded and more flexible role for the evaluator generally and to the study of values particularly.

In the often pictured input–output model of program functioning shown in Figure 3.2, the implicit assumption is that the part of the world depicted by the model is an entity unto itself. The universe we are concerned with supposedly starts with the input arrow and stops with the output arrow. This conceptualization fosters the utilization of traditional statistical design as this model rather seductively suggests that we are dealing with a cause-and-effect world. Like decision-oriented evaluation, this model greatly oversimplifies the entities at hand and totally ignores the values that may exist between relevant entities and the program.

Cooley and Lohnes's (1976) rediscovery and updating of Dewey's principles of valuation represents a significant addition to this input–output model. Primarily, it provides logical and theoretical justification for the study of values as part-and-parcel of the evaluative reform recently called for by other authors in the field. This justification has been sorely lacking. Unguided by a prudential theoretical basis, evaluation has been moved in directions not always conducive to the ultimate improvement of program settings or the planned change process.

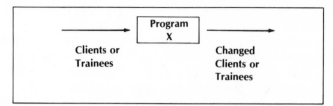

Figure 3.2. Input-output model of the domain of evaluation.

Another implication of this perspective on evaluation is that evaluation is not to be conceptualized as a single product (usually a monograph) delivered at the conclusion of an evaluation. Rather it is viewed as a process in which the evaluator interacts with all interested parties for an extended period of time, as called for by many naturalistic and qualitative models. This allows for the determination of values and the resolution of differences in opinion, viewpoint, and interests. Cooley and Lohnes consider this version of evaluation as "a process of conflict resolution through intelligent social deliberation [p. 16]." This approach suggests an interactive mode that allows the emergence of a common conceptualization of the program among all involved parties and fosters a consensus of program need, design, implementation, and evaluation.

This perspective also stresses the education of all persons involved in evaluation. Put simply, the evaluation should be a learning experience for all involved. It is not unrealistic to expect that the various parties to a program should come to understand more precisely what they are trying to do and why, how their programs achieve the results they do, and how each participant may individually facilitate the attainment of successively higher-level ends in a meaningful way. Thus, evaluation may be viewed as a procedure (or means) itself that has as its potential ends-in-view a more harmonious, pleasant and effective program setting.

A similar viewpoint is expressed by Cronbach et al. (1980). They compare the role of an evaluator to that of an educator and cite similarities between the two. According to this view, successful evaluation is one that helps evaluation consumers to deepen their understanding of the program being evaluated as well as the societal and organizational contexts in which the program operates. Similar to Parlett and Hamilton (1976), these writers stress the evaluator's role as an illuminator of program issues.

Finally, the value-oriented approach to evaluation is inherently humanistic in that it focuses on the total effects of a program and short- and long-range outcomes as part of a larger means-end continuum. Also emphasis is placed on the empirical validation of goals and values, thus preventing them from being determined arbitrarily. The conceptualization of a means-end continuum provides a foresightful vision of ultimate program effects. The goal-free bias inherent in the approach provides a rationale for being sensitive to unknown or unintended program effects. This theoretical view of evaluation has the potential for breaking the traditional mental set of evaluation and provides evaluators with a framework for providing useful information that goes beyond serving a particular decision-oriented context. Cooley and Lohnes (1976) capture the essence of the value-oriented perspective when they state,

> What has been missing in controversies over the schools is convincing evidence which relates choices of educational practices to ends which society values, ends which satisfy needs. Generating such evidence is what evaluation is all about [p. 18].

4

Some Systems Notions
and Concepts

The systems literature is both diverse and amorphous, spanning several disciplines. Each of the commonly used terms *systems theory, systems approach,* and *systems thinking* has many referents. The interdisciplinary nature of systems theory suggests its potential applicability to complex social problems. At the same time, systems terminology and "systems approaches" have been used so loosely that this potential is far from being realized:

> By now the term *systems approach* has so many references that it is practically impossible to define it except by pointing. This situation exists in part because systems approaches have been applied in so many disciplines and in part because the term has become prestigious jargon, an honorific, sales-boosting, argument-ending, grant getting magic incantation [Hoetker, 1972, p. 1].

It is our contention that the problem of too many referents leads to misunderstandings, criticisms, and emotionality which have hindered the extraction of anything practical and useful from the systems literature. Our hope in this chapter is to convey to the reader the breadth of the theory and procedure that has fallen within the rubric of systems.

It should not be surprising, then, that this chapter is a mixture of many ideas and concepts. We will begin with a discussion of some theoretical notions that have either explicitly or implicitly guided much of what is today called systems approaches and systems thinking. *General systems theory* represents a theoretical–philosophical stance in the sciences, whereas the oft-heard terms *systems approach* and *systems thinking* represent attempts to apply these theoretical notions to practical problems. Following a discussion of more theoretical systems ideas, we will review some applied systems notions and concepts.

General Systems Theory

Classical general systems theory can be traced to the writings of Ludwig von Bertalanffy. His initial articles on the subject appeared in the 1930s, but his writing did not attract much attention until after World War II. As a biologist, Bertalanffy began to characterize the properties of systems. Obviously influenced by his discipline, Bertalanffy wrote of living and nonliving systems, open and closed systems, static and dynamic systems, and adaptive and nonadaptive systems. His writings jelled into an area of inquiry now called applied general systems theory, which has spawned a proliferation of books, journals, special interest groups, and personalities.

The writings of systems theorists (Bertalanffy, 1950; Beishon and Peters, 1976; Boulding, 1956; Buckley, 1968; Churchman, 1968b; DeGreene, 1976; Emery, 1969; Laszlo, 1972; Rapoport, 1968a; van Gigch, 1978) provide convincing arguments that systems and the study of systems are necessary in order to understand man's increasingly complex interaction with man and with his environment. Increasing awareness of the multidimensionality in our world has led to the development of some methodological approaches by which we can understand complex events in larger contexts or expanded time frames. The sciences, particularly the behavioral and social sciences, have studied man's relationship to man and man's relationship to the environment in overly simplistic ways. Recent advances in these and related fields make it increasingly difficult to understand important concepts and principles without viewing the complex whole of which they are a part. The social and behavioral sciences have, in a manner of speaking, run out of simple problems to solve. Or, more correctly, they have found simple solutions to modern-day problems inadequate in the context of recent discoveries and advancements that have all but nullified many simple solutions of the past. Complexity is a fact of life, and complex problem solving techniques (and questions) that recognize a multidimensional world are in increasing demand. Simplistic portrayals of educational and human service programs quickly lose their credibility in the context of this complexity.

General systems theory represents an alternative or supplemental paradigm for understanding man and his environment. While traditional scientific inquiry attempts to understand man-within-his-environment by isolating the effects of single variables or small groups of variables and holding everything else constant, the general systems paradigm posits that it is impossible to understand complex events by reducing them to their individual elements.

The Gestalt school of psychology made popular the notion that the whole is more than the sum of its parts and that to understand the whole one must study the whole. This notion became an integral part of Kurt Lewin's (1951) field theory. Lewin provided a rationale for his theory by remarking that "Observation of social behavior is usually of little value if it doesn't include an adequate description of the character of the social atmosphere or *the larger unit of activity* within which the specific social act occurs [cited in Sutherland, 1973, p. 37]."

This *field view* is pervasive in systems theory, in that any whole is seen as consisting of subsystems. Any particular system is viewed as a subsystem within a larger system which, in turn, is itself a subsystem within some still larger system. Thus, within this view, any particular entity can be seen as a system unto itself, a whole that can be "broken down" into its component parts. This decomposition represents an *interoceptive view* of an entity. Simultaneously, the entity being examined can be viewed as a component part of some larger entity or process. Viewing this entity within "the larger scheme of things" represents an *exteroceptive view* of that which is being studied.

The interrelatedness of component parts within a larger system implies that change in an isolated part of a system will evidence itself throughout the system of which the altered component is a part. Thus to understand the implications and ramifications of any action, one may need to look beyond the immediate context in which the action occurs. General systems theory holds that the complexity inherent in this interrelatedness between systems cannot be understood through analytic reduction and experimental control. Instead, systems are viewed as organismic entities that live and die, evolve and decay. All systems are ultimately seen as interrelated, and changes in any particular system have implications throughout the larger system of which it is a part. Thus, there is no such thing as an "independent" variable; everything is ultimately related to everything else. This viewpoint differs considerably from traditional experimental design in which the immediate context is the focus of inquiry and great care is taken to control all "extraneous" variables.

In the more mechanistic conceptualizations of the world (in which parts are considered in isolation), the relationship between parts is generally viewed as additive or, at best, multiplicative, taking into account two or three variables at a time. This penchant is most apparent in the statistical models often used to characterize behavioral data. These models are often linear, assume that data points are independent, and are additive such that the whole is merely the sum of its parts (as when we partition *sums of squares* into main effects, interaction effect(s), and residual or error in analysis of variance).

A systems model, on the other hand, assumes that observations are dependent and that even the relationships between variables, much less the effects of a single variable, cannot be isolated at some point in space and time. All of these entities (or variables) are viewed organismically (i.e., they are considered to be in a constant state of flux, change, or evolution as they play out their role in some larger context). A systems view values the information gained from trying to understand this dynamic interaction. Whereas our traditional experimental methodology ignores this dynamism, usually through overt control (i.e., randomization, matched groups, or covariance procedures), a systems perspective focuses on this dynamism as the proper subject of inquiry in trying to understand what is actually occurring. In this view, the act of isolating entities at some point in space and characterizing the interrelationships with some simple additive mathematical model is viewed as primitive.

Another basic tenet of general systems theory is that there are isomorphisms in the systems studied by various disciplines. These isomorphisms are general characteristics or behaviors exhibited by *all* systems, regardless of their discipline of origin, be it a biological system, educational system, or air conditioning system. A major objective of this general systems philosophy is the search for similarities in structure and process occurring in the systems of different disciplines. For example, feedback and self-correcting behavior are likely to play a pivotal role regardless of the system under study. Similarly, most systems receive some kind of input, do something to that input to change or transform it in some way, and then return this changed input to the environment (usually called output). Furthermore, this processing by the system is *always* constrained in some way. The illustration in Figure 4.1a is a popular characterization of this process. Figures 4.1b through 4.1e present some simple examples of these three isomorphisms.

Techniques for understanding systems represent another class of isomorphisms. Van Gigch (1978) provides the example of the use of Markov chains as a means of describing the stages of breakdown and repair of large construction equipment, the offenses committed by recidivists, and the "brand-switching" behavior of shoppers. While we can probably safely assume that the behaviors of these machines, legal offenders, and shoppers are unrelated, somehow they are enough alike that we can characterize them all with a similar mathematical model. This gets at the heart of the concept of isomorphism: systems may be totally unrelated yet be isomorphic with each other on some continuum.

We will not belabor the point with additional examples of isomorphisms here. The point will be made subsequently that *programs* are best viewed as systems that share isomorphisms with all systems. General systems theory attempts to point out this generality in the way systems across disciplines are structured and how they behave, respond, and adapt in their environments. We will not present or use much of the notation and terminology developed in attempts to achieve this generality; it is too esoteric for our purposes.

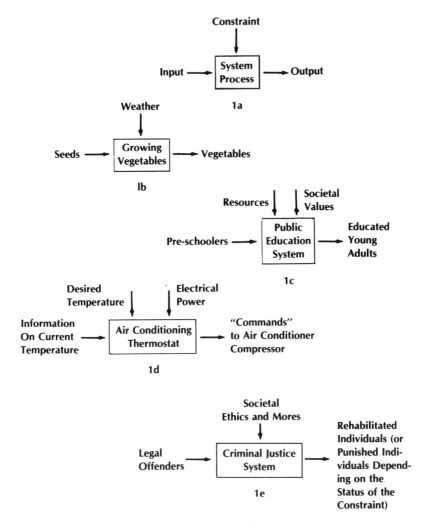

Figure 4.1. **Varieties of systems.**

Rather we want to communicate this view of the world, which emphasizes generality and relatedness. We ultimately believe that understanding the generality and relatedness of systems makes one a better evaluator, planner, and administrator.

General systems theory has generated a vast and amorphous literature. However, the basic notions of this theory can be reduced to the view of the world we have discussed. The question for our purposes is "Is it helpful to adopt this perspective when thinking about programs?" We will pursue this question in the following chapters. Our intent in this chapter is to review a collection of applied systems strategies that convey the breadth of activities that fall under the rubric of systems thinking.

The foregoing discussion of "classical" systems theory represents just one side of the systems coin. While the classical school of systems theory was developing a philosophical foundation for its thinking, some individuals in various fields were applying what were called *systems approaches* to real world problems. While we are indebted to the classical systems theorists for an orientation to the philosophy of systems, methodological tools and techniques have largely been the legacy of attempted systems approaches to technical problems. Attempts to apply systems concepts to real world problems have been made in many different disciplines. As will be pointed out, these activities have largely been performed independently of each other. Furthermore, some have evolved quite independently (or naively) of general systems theory. Despite the independence of these activities, however, some common threads run through them all, far too common and consistent a thread to be coincidental. A brief review of some of these systems approaches follows.

Systems Approaches to Education and Training

One generic class of systems approaches are step-by-step procedures utilized in the design, development, and implementation of programs, usually training programs. Conceptual models interconnecting the previously independently conceptualized planning, development, and evaluation processes appeared as early as the 1920s. (See Silvern, 1971, for a review of these developments.) Initially utilized in vocational education and military training, the use of instructional systems development (or systems approaches to training) has mushroomed in almost all areas of training and education. Montemerlo and Tennyson (1976) found in the literature over a hundred separate models or procedures for developing instruction.

These models were unique in their efforts to include within a single domain methodologies to actually improve the process by which programs were being planned, developed, *and* evaluated. To accomplish this purpose various systematic approaches to instructional development were introduced posing "front end" or predevelopment tasks for the evaluator, thus unifying and integrating the previously separate processes of program planning, development, and evaluation.

Kaufman (1972), in the first modern text dealing with educational planning from a systems perspective, defined *system* as "The sum total of parts working independently and working together to achieve required results or outcomes, based on needs [p. 1]."

He defined the *systems approach* as "A process by which needs are identified, problems selected, requirements for problem solution are identified, solutions are chosen from alternatives, methods and means are obtained and implemented, results are evaluated, and required revisions to all or part of the system are made so that the needs are eliminated [p. 2]." The particular systems approach advocated by Kaufman represents a type of logical problem solving for identifying and resolving educational problems.

One example of the systems approach applied to planning and evaluation is the *Interservice Procedures for Instructional Systems Development* (US Army Training and Doctrine Command, 1975), a five-volume monograph on the "how to" aspects of instructional systems development. Although developed for the military, this work represents a broad application of the systems approach to training useful in virtually any type of setting. The Interservice Procedures are divided into five separate and distinct phases to be carried out successively. These phases as described in the executive summary of the project are the following:

1. *Phase I, Analyze.* This phase deals with procedures for defining behaviors, breaking these into statements of tasks, and using numerical techniques to combine the best judgment of experienced professionals to select tasks for training. Phase I also presents processes for constructing performance measures and the sharing of occupational and training information within and among client groups. It provides a rationale for deciding whether tasks should be trained in schools, on the job, or elsewhere, and also requires consideration of the interaction between training and job performance.

2. *Phase II, Design.* This phase deals with the design aspects of the training program within selected settings. Design is considered in the architectural sense in which the form and specifications for training are laid down in careful detail. Phase II reviews the considerations relating to entry behavior of two separate kinds: general ability and prior experience. A rationale is presented for establishing requirements based on the realistic evaluation of both of these factors.

3. *Phase III, Development.* This phase refers to the actual preparation of instruction. Determinations are made about how the students will be managed, the kinds of learning experiences they will have, the activities in which they will be engaged, and the form and content of the instructional delivery system. Techniques are presented for the careful review and adaptation of existing materials. Procedures for the systematic design of instruction that can be delivered in a variety of media are also included. Phase III concludes

with a procedure for testing and evaluating the instruction to insure that its performance meets expectations.

4. *Phase IV, Implementation.* This phase treats the necessary steps to implement the instruction according to the plan developed in Phase III. Two steps highlight Phase IV, that of training the staff in the procedures and problems unique to the specific instruction, and actually bringing the instruction on-line and operating it. The Phase IV effort continues as long as there is a need for the instruction.

5. *Phase V, Control.* This phase deals with procedures and techniques for maintaining instructional quality control standards and for providing data from internal and external sources upon which revisions can be based. Data collection, evaluation of the data, and decision making about the implications of the data represent the three principal functions described in Phase V. Emphasis is placed on the importance of determining whether the trainees are learning what was intended, and upon determining whether what they have learned is of benefit in carrying out post-training responsibilities.

These phases describe the functions necessary to analyze instructional needs; to design, develop, and implement instruction; and to maintain quality control of instruction. Of primary importance is the sequential relationship of functions within and between these phases.

In a similar manner, Dick and Carey (1978) integrate the processes of planning, development, and evaluation into a 10-step approach. Their steps are: identifying instructional goals, conducting an instructional analysis, identifying entry behaviors and characteristics, writing performance objectives, developing criterion referenced tests, developing an instructional strategy, developing and selecting instruction, designing and conducting formative evaluation, revising instruction, and conducting summative evaluation. Their procedure is described in some 200 pages and 10 chapters explicating each of these processes and integrating them into a single model.

But these are only a few examples of the systems approaches that have blossomed in education and training. Other approaches have required the evaluator or developer to conduct needs assessments, prepare program specifications, perform task and learner analyses, and identify human and material resources available for the development task.

A systems approach is generally employed for the purpose of assuring that the program does what it is supposed to do. Accordingly, a systems approach to training may specify procedures for establishing the accuracy and representativeness of the objectives upon which a program is to be based, for analyzing the characteristics of learners and the learning task, and for monitoring the development process itself. As can be noted from the previous examples these responsibilities have blended together the fields of instructional development and evaluation into a single framework. However, this unifying framework containing two formerly distinct areas of specialization has not been without its problems.

A major question around which some concern has been raised is whether the evaluator should be distinct from the developer or whether these roles represent responsibilities that can be filled by the same individual working within the context of a systems approach. Some authors (e.g., Scriven, 1967) have warned that when role distinctions become unclear, the program may suffer from what has come to be called co-option. This refers to the situation in which the evaluator is so immersed in the values, feelings, and intents of the developer that evaluations are no longer an objective guide to program effectiveness. On the other hand, some authors (Butman and Fletcher, 1974; Grobman, 1968) have contended that development of an education or training program is so closely tied to evaluation that any separation of roles or functions is at best an artificial distinction that may detract from rather than add to the development process. To some extent, the use of "third party" (or independent) summative evaluations has dissipated differences between these perspectives.

Although little has been written about the role and function of the evaluator within the context of program planning and development, it is not uncommon for a program to be planned and developed in such a way as to either encourage or preclude a certain kind of evaluation; or, once the program has been developed, for the evaluator to be forced to take a certain approach to evaluation, regardless of its responsiveness to clients' needs or the original intent of the evaluation. As we suggested in Chapter 3, the evaluator can serve critical functions early in the development process to prevent just such an eventuality. For a more thorough treatment of the systems approach in education and training see Banathy, 1968; Briggs, 1977; and Davis, Alexander, and Yelon, 1975.

Systems Approaches to Management Science

Another area of systems technology that emerged during the post-World War II years was in management science. This class of systems approaches represents attempts to improve organizational design, efficiency, effectiveness, and accountability through implementation of a prespecified managerial plan. Most of the literature and process of management science is viewed as constituting a "systems approach" for several reasons. Many of these approaches such as Management by Objectives—MBO (Drucker, 1964, 1974); Planning-Programming-Budgeting Systems—PPBS (Hinricks, 1969); Goal-Attainment Scaling—GAS (Kiresuk and Lund, 1978; Kiresuk and Sherman, 1968); and the Program Evaluation and Review Technique—PERT (Paige, 1963; Tosi and Carroll, 1976) explicitly claim a systems perspective in their approach. These approaches represent attempts to deal with a program at the management level. In other words, the program is viewed as a functional whole consisting of component parts. Concomitant with this view is an awareness of the environment in which the program operates. These approaches emphasize interaction of component parts and the meshing of results or outcomes with general system objectives. Thus, concepts such as *interface* and *hierarchy* are particularly relevant to these management approaches. Some of these management perspectives are reviewed below.

OPEN SYSTEMS PERSPECTIVES ON ORGANIZATIONS

Baker (1973), Drucker (1974), and Osborn and Hunt (1974) provide convincing arguments that an organization should be viewed as an open system operating within the context of some larger environment. This perspective shifts management's attention towards consideration of the organization's external as well as internal affairs. Such a perspective gives management a conceptual basis for monitoring ever-changing external demands and contingencies while simultaneously attending to internal stability, efficiency, and productivity. In this way the internal workings of the organization or system can be manipulated through management to produce outcomes consistent with external demands and pressures. Classical management strategies have traditionally attempted to foster management capability through decomposition of a complex organization into component parts, simplifying the enormity of an organization or its tasks. Usually this decomposition was in terms of functional units (production, marketing, and the like). Each component could then be further subdivided to the level of detail needed for effective management. While this did simplify complex organizational structures, the exclusive internal focus precluded consideration of highly relevant considerations of the connection between the organization and its environment. Thus the focus was highly internal at the cost of obfuscating critical environmental events that could impact internal management decisions.

Forrester (1965, 1975) has presented some guidelines for the development of organizations of the future which have come to be viewed as classic in the field of organizational theory and design. His design principles are based to an extent on social–psychological principles such as decreasing authoritarianism in favor of more democratic forms of control and decision making. More central to this argument is his emphasis on a thorough understanding of systems dynamics as the basis for the development of policy which, in turn, guides all subsequent organizational activity. Forrester also stressed the use of modern technology, particularly computers, to maximize information flow within and outside the organization.

Simon (1973) gives particular emphasis to information processing and decision making in the design of organizations, citing these as particularly relevant criteria for the design of organizations whose primary product is service rather than goods. The thrust of Simon's argument is that it is attention, not information, that is a scarce resource for modern man and that our organizational structure should be designed so as to best utilize management's attention to internal as well as extra-organizational information.

PROGRAM PLANNING AND BUDGETING SYSTEMS (PPBS)

PPBS (also called Planning–Programming–Budgeting System) evolved in the federal bureaucracy and was first applied on a large scale in the Department of Defense in the early 1960s under the direction of Defense Secretary Robert McNamara. This management process was extensively used in all agencies of the executive branch of government during the Johnson Administration. While this system ultimately received much criticism at the federal level for political, social–psychological, and implementation/preparation reasons, as well as from genuine weaknesses in the technique, this systems-oriented management approach represented the first broad-scale effort in the area. It is still being implemented in parts, especially by nonprofit agencies (Sorensen and Elpers, 1978), and has served as a model or guideline for the development of related management strategies such as Management By Objectives (MBO) and human resources accounting.

PPBS is essentially a four-step procedure in which the organization's (or organization subunit's) objectives are first specified and analyzed for relevance and appropriateness. Then any particular program output is analyzed in terms of the objectives elucidated in Step 1 to determine whether the objectives are being met. In the third step *all* costs of the program are measured, including as-yet unincurred but anticipated future costs. In the final step, available alternatives are evaluated in terms of meeting basic objectives at the most advantageous cost.

This approach has several limitations. For example, if a mediocre job is done in Step 1 (for which there are no criteria for acceptable performance) or if all costs are not fully and completely anticipated in Step 3, then the results of this management procedure can be inadequate. Thus PPBS is not an algorithm or lock-step procedure which ensures the "right answer" every time, but rather is a heuristic to guide and aid thinking about a problem.

OTHER SPECIFIC MANAGEMENT APPROACHES

MBO (Drucker, 1954, 1964, 1974) is similar to PPBS as a management strategy in that it is a stepwise procedure that begins with an elucidation of broad aims or goals. In MBO, goals are distinguished from objectives which, in turn, are distinguished from measures. The relationships between goals, objectives, and measures are viewed as hierarchical and all three are specified prior to the beginning of an organization's or individual's activity. The *goal* is generally some statement about a desired state of affairs. Once this desirable state has been elucidated, the goal is translated into a set of *objectives*, which constitute interpretations of the goal. Objectives should specify the time dimension, target groups, variables of interest, change expected (kind and degree), and the activity to be employed to meet the objective. *Measures* then specify the operational definitions of the objectives (i.e., the specific numerical or qualitative evidence that the objective was or was not achieved).

Specifying goals, objectives, and measures ahead of time both guides the individual's or organization's activities *and* unambiguously elucidates how the evaluation will be conducted. As with PPBS, it is critical to MBO that goals be set realistically (if initial goals are unattainable, the MBO approach to management would be self-defeating) and honestly (if easily attainable goals are specified, then the MBO approach would have success built into the procedure regardless of what actually happened). The interrelationships between goals, objectives, and measures have to be consistent and logical.

MBO is but one specific management strategy related to PPBS. Another highly similar approach, often utilized in mental health evaluation, is Goal Attainment Scaling—GAS (Kiresuk and Lund, 1978; Kiresuk and Sherman, 1968; Sherman, 1977). Goal Attainment Scaling contains the additional wrinkles of prioritizing goals and prespecifying varying degrees of success or failure in attaining each specific objective.

Another cluster of management approaches closely related to PPBS (and MBO and GAS) is cost-benefit analysis. Also related are cost accounting, cost-effectiveness, and output value analysis. The cost-effectiveness approach attempts to derive meaningful cost outcomes (x amount of benefit per unit cost) for each of several program alternatives to guide subsequent managerial decision making. While this is virtually identical to Step 4 of PPBS, considerable effort has been devoted to the conceptualization and development of methodology to more accurately specify both outcomes and costs, and methods of linking the two (cf. Halpern, 1977). Sorensen and Grove (1978) provide an excellent review of these cost-effectiveness approaches.

Related to the specific management approaches discussed above are issues surrounding the development and use of management information systems (MIS). Each of the management strategies discussed contains within it certain prioritized information needs. With the recent availability of highly sophisticated computer support systems capable of flooding management with literally reams of information, the development of management information systems has closely paralleled the evolution of management strategies. As Simon (1973) has pointed out, program managers are faced with a "rich soup of information" and computer support services and well-conceived organizational structures can play an indispensable role in focusing managers' attention efficiently. (See Elpers and Chapman, 1978; and Sorensen and Elpers, 1978, for papers on MIS design for evaluators.)

When one uses the term *systems approach,* some listeners will have as their referent some variant of one of the management perspectives or management strategies discussed above. It should be becoming clear to the reader that the number of "things" that can be termed systems approaches is indeed large. At the risk of belaboring the point however, one other kind of systems approach will be examined here, both because of its substance as a systems approach and because it will present some concepts useful to us subsequently. This is the area of systems analysis.

Systems Analysis

The term *systems analysis* has about as many referents as does the term *systems approach* and, indeed, much of the foregoing, presented as examples of systems approaches, can be and have been called types of systems analysis. However, as Coursey (1977) points out, systems analysis refers to a whole family of techniques for analyzing systems, designing systems, and for simulating and testing systems. Many disciplines have several methodologies for doing systems analysis, varying with viewpoint and purpose. These procedures are far too numerous to review here. Rather, we will discuss some common properties of these varied techniques and present a few selected examples.

PROCESS MODELS

A common type of systems analysis is a lock-step procedure utilized to attain some prespecified purpose. These approaches are similar to the con-

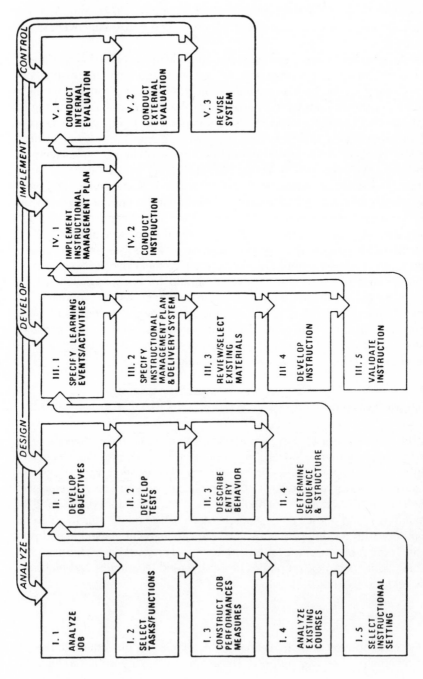

Figure 4.2. Interservice Procedures for Instructional Systems Development Model (Branson, 1978).

cept of *algorithm;* however, rather than guarantee a correct (and reliable) result, these procedural "roadmaps" lay out the course of action to be followed to obtain the most optimal solution possible in a given situation by focusing resources on relevant tasks. The *Interservice Procedures for Instructional Systems Development,* cited earlier in this chapter, is an example of this variety of systems analysis. A schematic diagram of this procedure is presented in Figure 4.2.

The process depicted in Figure 4.2 is a generic one, developed for implementation in a variety of environments. Other similar stepwise procedures have been developed for specific tasks. These process models are often put on a time-line for purposes of scheduling and planning. The Program Evaluation and Review Technique (PERT) and the Critical Path Method (CPM) are among the more well-known of these network scheduling techniques. The ultimate goals of these procedures are usually concerned with optimizing the use of resources and economizing the time needed to complete a project.

FORECASTING MODELS

Several models have been developed in specific fields to predict the likelihood of certain events under varying environmental conditions (Forrester, 1975; Meadows, Meadows, Randers, and Behrens, 1972). Some of these are mathematical (most often linear) models that specify the probability of some future event on the basis of current and past events and conditions. However, not all future forecasting techniques are linear, or even mathematical. These techniques have as their objective one or more of the following:

1. Generation and consensus of issues: to generate or obtain consensus on concerns and questions related to an existing or future program
2. Relationships: to show the relationship among entities that impinge upon a program as they might occur in some existing or future setting
3. Generation of alternatives: to generate alternative possibilities or courses of action for use in planning a program
4. Simulation: to simulate or "act out" some existing or future condition that can impact a program

Some of the more popular forecasting techniques are described below.

Force analysis. Force analysis (Culbertson, 1969) employs the construct of *societal-forces-at-work* in predicting future states. A force is defined as a set of events, pressures, or technologies whose impact causes specific changes in the environment in which a planned program is to operate. Knowledgeable observers, called analysts, scan the present and immediate

past environment for forces that can impact the program under consideration. Reliance is placed on descriptive and/or analytical formulations, including those that might be produced by questionnaires, interviews, and
literature reviews. After selecting a manageable number of force descriptions
that describe the past nature or previous impact of these forces on similar
programs and environments, analysts predict the future nature of each force
and its probable impact on the proposed program. These force descriptions
are created from leading questions given to each force analyst. Typical
questions (from Haskew, 1974) used to guide the force analyst are

1. What are the salient features of the force (including trends over time?)
2. Are there significant and identifiable factors that have made the force
   especially visible during the last decade?
3. Are there relationships between the force and dimensions of its environment that have special implications for assessing impact on the
   program?
4. Has the force had demonstrable impact on similar or related programs?
5. What is the essential mode of impact or influence of the force?
6. At what points in the system has the force had greatest influence?
7. What are some major unresolved and emerging issues posed by the
   force?

Delphi technique. Delphi (Helmer, 1967) is a procedure for eliciting and
refining the opinions of a group of informed individuals without the scheduling restrictions normally imposed by having the group meet. The Delphi
technique relies on informed judgment and anonymity through the use of
questionnaires to achieve consensus without asking the group to arrive at a
common opinion or to meet in face-to-face confrontation.

Application of the Delphi procedure consists of asking a panel of experts
to respond to a series of questionnaires on which they make independent
judgments about a selected topic. The Delphi procedure consists of the
following six steps:

1. An initial questionnaire is developed and administered that elicits
   opinions about a key issue or set of related issues.
2. From an analysis of the results, a second questionnaire is prepared that
   contains the items on the original questionnaire plus a statistical summary of the distribution of responses obtained from the initial round.
3. Panel members who have given the most extreme responses are asked
   to reconsider their responses and to change their responses if they so
   desire.
4. Respondents who decide to change a response from the previous
   round are asked to provide reasons for doing so.

5. In subsequent rounds, responses are statistically summarized and a list is provided of the reasons given by panel members who have changed responses.

6. This process is continued until panel members no longer wish to change responses or a pre-established number of rounds is reached.

Decision matrices. A decision matrix (Ayres, 1969) is a means of detecting and examining relationships between any given set of factors. A decision matrix is used to order and link units of information so that the consequences and implications of their relationships can be judged. The technique generally employs a two-factor matrix: one factor listing data (e.g., goals, events, or content domains) vertically and another factor listing a different set of data (e.g., possible courses of action) horizontally. Numbers between 1 and 0 are placed in each cell of the matrix, indicating the estimated strength of the relationship of each pair of items constituting a row and column intersection. For example, by defining column entries as desirable program goals and row entries as alternative courses of action, the "goodness of fit" between stated goals and various alternative programs can be determined.

Cross impact analysis. Cross impact analysis (Rochberg, Gordon, and Helmer, 1970) is a procedure that adjusts the probability that certain hypothesized effects for a program will occur, in view of judgments of the potential interactions between the program and events relating to it. The primary goal of cross impact analysis is to systematically structure the interactive relationships between and among possible events so that program planners can consider both the probability of occurrence of an event critical to program functioning and the impact this probable occurrence could have on other events (e.g., program outcomes). In this manner, relationships between and among events related to the program are clarified and possible implications of alternative program activities are considered. The magnitude of the effect of one event (E) on another is shown graphically by the size of arrows placed on each cell of the matrix, or statistically by inserting an estimate of the percent of change in one event produced by another event. The matrix allows the analyst to systematically manipulate individual events while monitoring how the probabilities of these events will affect all other relationships under consideration. An example of a cross impact matrix is shown in Figure 4.3.

Relevance trees. Relevance trees (Martino, 1972) are logically derived hierarchies of goals, events, or content domains from which a program can be planned. The purpose of a relevance tree is to set out all possibilities that may be relevant to the planning of a program from which selections for its actual design and implementation can be made. Relevance trees have three

	Then the probability of these outcomes (O) would change as indicated.			
If this event (E) were to occur	O_1	O_2	O_3	O_4
E_1 Resource allocation level is reduced.		↑		↓
E_2 Only a single delivery system is used.				↑
E_3 Client characteristics shift to reflect population trends.	↓			↓
E_4 Program personnel fail to receive regular inservice training.	↓			

Figure 4.3. A cross impact analysis showing the interactive effects of four events.

distinct features: (a) branches representing program goals or content domains extend downward and represent in tree-like fashion a closed set of alternatives, that is, an exhaustive list of all relevant options for that goal or content domain, (b) the branches extending downward from any goal or content domain represent mutually exclusive program alternatives, methods or strategies, with nonoverlapping characteristics, and (c) each program goal or content domain is completely described by all branches below it. The requirements that branches represent closed sets and be mutually exclusive provide the means whereby relevance trees display alternatives for program design. Each node or junction within the "tree" can be viewed as a decision point for program planning, with each set of branches at a given node representing the alternatives available at that point (see Figure 4.4).

Morphological analysis. Morphology is the study of form and structure. Morphological analysis (Zwicky, 1969) is a method of inquiry that focuses on the systematic elucidation of all possible solutions to a well-stated problem without prejudging alternative courses of action. Morphological analysis consists of:

1. Identifying and concisely formulating the problem to be solved (i.e., determining the purpose or purposes for which a program might be planned)
2. Identifying all parameters that might be of importance for the design of a given program (i.e., identifying such variables as instructional context, mode of presentation, background and experience of partici-

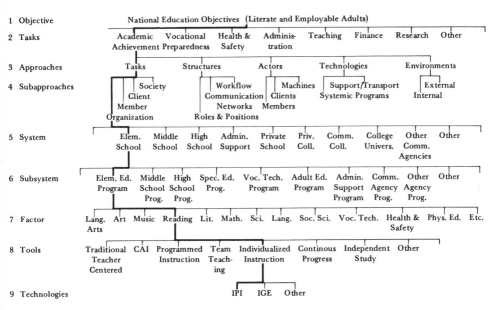

Figure 4.4. **Relevance tree applied for national education objectives (McGrath, 1974).**

pants), that may be structurally related, and from which a program design might be constructed

3. Constructing a multidimensional matrix that represents all possible permutations or alternative paths for the creation of a program. This involves stating parameters (P_1, P_2, . . .) and parameter values (P_1^1, P_1^2, . . .) in matrix form. For example,

$[P_1^1, P_1^2]$ (e.g., participant groups)

$[P_2^1, P_2^2, P_2^3]$ (e.g., program strategies)

$[P_3^1, P_3^2, P_3^3]$ (e.g., administrative organizations)

$[P_4^1, P_4^2, P_4^3, P_4^4]$ (e.g., delivery systems)

$[P_5^1, P_5^2]$ (e.g., content areas)

4. All solutions (paths depicted by the multidimensional matrix) are scrutinized and evaluated for internal contradictions that would result in meaningless solutions.

5. Solutions are selected and applied. This step involves making a determination of the desirability and relationship to program goals of all possible solutions that were not deemed contradictory in Step 4.

Scenarios, simulation, and gaming. A *scenario* (Sage and Chobot, 1974) is a well thought out story about some possible future state. Scenarios are used in program planning and forecasting to depict a particular set of conditions that might evolve and to provide a framework for possibly changing those conditions.

The primary goal of a scenario is the identification of causal relationships and key decision points from both quantitative data (e.g., empirical studies of needs and conditions) and qualitative data (e.g., testimonials from experts and clients). In selecting data for a scenario, greater emphasis is placed on the expertise, reputation, and credibility of individuals supplying the data than on the methods used for data collection and analysis.

Scenarios are closely related to both simulation and gaming. A *simulation* is an operational run-through of all or part of the program in a controlled but life-like setting typical of the environment in which the program will operate. *Gaming* is the abstracting of key elements of a program (e.g., its administrative structure, organization, or technology) and the integration of these elements in a simulated environment using rules that approximate those operative in the real world. Contrary to simulation and gaming, scenarios represent hypothetical and descriptive accounts of program operation that do not require the program or parts of it to be actually implemented.

Psychodrama/role-playing. Psychodrama (Mixon, 1972) is role-playing for the purpose of analyzing the effects on people of a carefully described situation. The underlying argument for the use of psychodrama in planning and forecasting is that planning and policy decisions, to be truly effective, must be tested with real people in situations that closely approximate those that would be created if the program was in operation.

Role-playing requires the participants to be engaged in either simulation or game-type activity complete with detailed instructions, scripts, or other materials necessary for replicating the real program. These instructions, scripts, and materials mirror the actual program in all relevant aspects, rather than simply asking participants to make judgments about an imaginary or hypothetical situation. Role-playing assumes that realism and spontaneity of speech and activity are possible when participants are introduced to a simulated or game-like environment approximating real-life contingencies.

Man has always been enchanted by the future, and the hope of controlling it, of course, predates the evolution of systems theory, classical or otherwise. However, it is undeniable that forecasting models and techniques embody many concepts integral to classical systems theory, reflecting a global perspective in considering problems and possible solutions. The area of futures forecasting is intimately tied to the field of planning. The point we want to make is that futures forecasting techniques are inherently systems-oriented methods. They are concerned with determining the potential effects of an intervention in an environment and thus focus exclusively on the systemic properties of the entities being planned and/or the wants and needs that are being planned for. Planning cannot occur in a vacuum, and the "stuff" of planning are such elusive phenomena as complex interactions and interdependencies among various "forces" in the environment. Planning techniques that help thinkers elucidate this "stuff" are far reaching in their implications and generally unburdened by assumptions about what can and cannot be, creating the possibility of generating novel alternatives to old problems. We merely want to make the point that when the term *systems approach* is used, some listeners may take some form of forecasting models and techniques as the referent. A summary of the advantages and limitations of the above forecasting techniques is presented in Table 4.1. (See also Hencley and Yates, 1974, for a review of these and other techniques.)

STAND-IN MODELS

While many systems analysis techniques may not share similarities with the forecasting models described, almost all systems analysis techniques utilize some simulation or, at least, a representation of the system or object of focus. In more technological applications, (e.g., man-machine systems), the models utilized tend to be hardware, often "mock-up" versions of the real thing, be it an air traffic controller's radar display or a weightlessness simulator used in spaceflight training. These models are indispensible in planning, training, and research.

In some disciplines, *graphic* stand-in models are usually of the input-process-output variety illustrated in Figure 4.1 and may share the common feature of presenting a program's content hierarchically. This graphical, hierarchical presentation of a highly complex program's internal structure and functioning is often used as a way to communicate with oneself and others when faced with a large amount of complexity within a system. Simultaneously, these graphic models may be used to show a program's juxtaposition with its environment. Such a view of the program might be used to illustrate organizational structure, or to describe interdependencies existing between component parts in some environment, or to examine the consistency between subcomponent outputs and larger system objectives. The idea that "a picture is worth a thousand words" illustrates the value of graphic stand-in models.

Graphic modeling methodologies have evolved in several different fields. In the field of training and education, Silvern (1969) has developed a Language for Optimizing Graphically Ordered Systems (LOGOS). LOGOS is a flowchart-like language for model-building that utilizes specific graphic conventions along with alphanumeric labels and narrative to convey a training program's content, structure, and process. The basic components of

Table 4.1
Summary of the Advantages and Limitations of Eight Forecasting Techniques

Primary goal	Techniques	Advantages	Limitations
Generation and consensus of issues	Force analysis	Can heighten sensitivity of decision makers to social, political, and technological forces at work in the program environment.	Judgments are subjective, requiring cross validation of points of of view rendered.
	Delphi	Can be used to obtain consensus on a wide range of program planning issues without the need for respondents to meet face-to-face.	Data may be distorted by the biased selection of panel members. An acceptable level of consensus may not be reached. The method itself may create as well as elicit opinions.
Relationships	Decision matrices	Can match various dimensions of a to-be-developed program with alternative courses of action to identify mutually supportive relationships.	Users may need training in quantifying relationships between different levels of the matrix. Skill and experience of users may influence quality of the output.
	Cross impact	Forces users to publicly state interactive effects of key events. Can identify alternative versions of a program from which a final version can be selected.	Should be used with other techniques to systematically determine relationships between events.
Generation of alternatives	Relevance trees	Can produce a framework for generating alternative courses of action. Can uncover new, previously unconsidered possibilities. Can explicate the dependency of "lower-level" concepts on "higher-level" concepts.	Cannot be certain an exhaustive list of possibilities has been generated at each node.

(Continued)

Table 4.1 (*Continued*)

Primary goal	Techniques	Advantages	Limitations
	Morphological analysis	Can be used to determine all possible solutions, minimizing the possibility of overlooking promising but unconventional solutions.	May not be possible to formulate program goals precisely. Number of programming alternatives generated may be excessive.
Simulation	Scenarios, simulations, and games	Can be used to specify the details and dynamics of potential problems and eventualities, which if ignored could create a "crisis."	Descriptions are generally oversimplifications of future states, requiring multiple trials or "runs" to maintain perspective.
	Role-playing	Can identify the unexpected and its significance for program operation. Through the preparation of instructions, scripts, and materials can identify the role-rule context implied by the program.	Can lead to highly idiosyncratic accounts of a program's performance if not repeated to obtain patterns and trends.

LOGOS are blocks indicating functions and *signal paths* (or arrows) indicating the flow of information or data.

According to Silvern (1969), "the thought expressed by a LOGOS flowchart is a conceptualization in the form of a graphic analog representing a real-life situation [p. 18]." The idea of a graphic display of a program is not unique to Silvern. Kaufman (1972) suggested the use of idiosyncratic flowchart-like models for purposes of program design and analysis. Silvern's unique contribution appears to be his attempt to standardize the graphic jargon of education and training.

While Silvern, Kaufman, and others have offered general modeling techniques for use in developing educational and training programs, their subject matter has been "soft" (i.e., loosely defined) enough that the degree of specificity required has not necessitated a more formal and rigorous methodology. Within the field of computer software design, however, the press for specificity has been much greater. Here such problems as matching a program or compiler's output and capability with the needs and wishes of user groups has traditionally been a thorny issue. In part, this problem is the result of the enormous amount of complexity inherent in many software applications. Examples include designing the software to control the flight of pilotless reconnaissance planes, designing telephone switching systems which simultaneously control all long distance calls across entire continents, or designing computer security systems for a large corporation utilizing thousands of remote access terminals. These problems require conceptual models encompassing high degrees of sophistication and specificity.

One of these design methodologies, which holds promise for program planning and evaluation, is the Structured Analysis and Design Technique—SADT (Ross, 1977; Ross and Schoman, 1977). SADT, depicted in Figure 4.5, is essentially a decomposition methodology that allows one to break a complex program down into its component parts while simultaneously holding the whole program in view.

SADT is a generic methodology that can be used to describe any kind of system with a high degree of specificity. Moreover, this technique is capable of describing both content and process simultaneously, something with

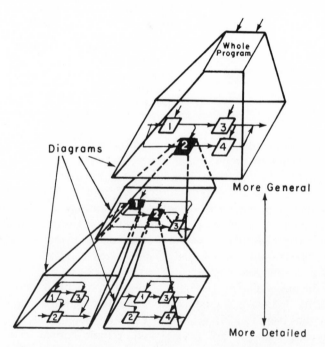

Figure 4.5. Structured decomposition. (This figure is reprinted from *Structured Analysis (SA): A Language For Communicating Ideas*, by D. Ross, copyright © 1977, IEEE Transactions on Software Engineering, SE-3(1), 16–34.)

which the LOGOS language has some difficulty. Lastly, SADT relies on fewer special graphic conventions than does LOGOS, making it more readable and easier to learn. SADT has appeared in several education and evaluation-oriented journals and can be used with any type of program (Borich, 1979; Borich and Brackett, 1978).

The following assumptions underlie the use of SADT in program evaluation.

1. Programs are best evaluated by first building a model that expresses an in-depth understanding of the program, sufficiently precise to serve as the basis for program development.
2. Analysis of any program should be top-down (moving from general to specific outcomes), modular (take into consideration all important components), and hierarchic (show how the components are tied together).
3. Program activities should be represented by a diagram which shows their interfaces and place in the hierarchic structure.
4. The model-building technique should represent behaviors to be produced, activities to be provided and the relationships among behaviors and activities.
5. All planning, design, development, and evaluation decisions should be in writing and available for open review by all team specialists.

SYSTEMS ANALYSIS SUMMARY

As Pfieffer (1968) points out, the use of a model, graphic or otherwise, greatly facilitates our ability to think about our evolving world:

Basic to the entire systems notion is the concept of a model, a simplified but controllable version of a real-world situation which serves a function roughly comparable to that of a laboratory experiment in the physical and biological sciences. More often than not it is impossible to analyze major problems without some kind of model. The problems are far too complex to cope with by ordinary inspection or common sense. We tend to approach phenomena in "linear" terms, as if events always followed one after the other in a direct chain reaction, like the falling of a row of dominoes. Our sentences and our ideas are organized that way.

We might have been able to depend on our own simple capacity for this kind of thinking in a simpler, slower-moving, and less closely-coupled world. But the problems confronting us today involve incredibly complex mazes or "nests" of interconnections and linkages rather than straightforward associations and cause-and-effect sequences, and the results of important changes may be extremely difficult to predict. In such cases a good model can help appreciably by supplementing intuition and judgment. It is one thing to rely solely on speculations about how people will behave under certain real-life circumstances, and something else again when one can objectively investigate possible behavior patterns by simulating those circumstances in a model [p. 6; this quote and subsequent quotations cited to Pfieffer (1968) are reprinted from *New Look At Education*, by J. Pfieffer, copyright © 1968, by permission of Odyssey Press].

Summary

The methodology and undergirding philosophy (explicit or implicit) of these systems approaches varies considerably. They have evolved in different disciplines in response to different problems and presses. Yet all are similar in their emphasis on describing the systemic qualities of a phenomenon. Halpert (1970) succinctly summarizes this common feature of these various examples of systems thinking:

A systems approach is a logical way of examining and trying to solve problems. It attempts to "map the territory" to show inter-relationships among elements in the system (or organization), and to identify operations basic to the mission of the organization. The systems approach stresses decision points and relations among various functions of the organization [p. 51; this quote and subsequent quotations cited to Coursey (1977) are reprinted from Coursey, R. D., An Overview of techniques and models of program evaluation, in R. D. Coursey (Ed.), *Program Evaluation For Mental Health Methods, Strategies, Participants*, by permission of Grune and Stratton, Inc., 1977].

The foregoing discussion points to four potential cradles from which any particular systems approach may arise: (a) classical systems theory, (b) the instructional system development movement, (c) management science, and (d) systems analysis. Unfortunately, even here our list is not exhaustive. Systems approaches with varying degrees of similarity to those already described also exist in such fields as industrial psychology, operations research, ethnography, transportation and traffic management, and economics, to name only a few.

The "roots" or "common cradles" of systems thinking we have presented could be conceptualized quite differently. For example Lilienfeld (1978) cites three distinct sources of systems theory that include most of what we have designated here as systems approaches or conceptualizations:

(1) the engineers, the cyberneticians, and the operations researchers, located for the most part in the aerospace industry and in the sectors of the academic world that are oriented toward technology; (2) academicians, including economists, political scientists, sociologists and biologists, oriented primarily toward science or "scientism" in one form or another; (3) the bureaucrats of the social service world [p. 268].

Although we find Lilienfeld's classification incomplete and resembling an attack on systems thinking and thinkers, it is presented to show that other conceptualizations of systems thinking and its origins can be found in the literature. While the taxonomy we have presented is likewise spotty and somewhat arbitrary, we feel it is a useful introduction to the breadth of ideas, philosophies, and procedures that are commonly considered as the systems approach.

The significance of systems thinking for program evaluation is that it forces the evaluator to view programs as open systems, that is, as programs that have permeable and flexible boundaries. As open systems, programs are in continual interaction with their environment, changing this environment as well as being changed by it. One type of relationship that programs can have with their environment is manifest in the means-end continuum that links the program to some higher-order program or system from which it derives its purpose. A systems perspective can be an indispensible tool for elucidating this purpose and, in the process, delineating the criteria by which a program's merit or worth can be determined.

Also consistent with a systems orientation is the view that individual programs are parts of larger programs and, therefore, can be meaningful only in relation to this larger program. This view is diametrically opposed to the reductionistic or incremental approach to evaluation in which the evaluator focuses on parts of the program or the whole program but not the relationship between the program and the larger context to which it is expected to contribute. A systems orientation stresses the importance of evaluating a program in relation to the context or systems in which it functions and emphasizes the pitfalls of any conceptualization that prevents one from seeing a program as part of some larger context.

Lastly, although many different techniques have been included in this review of systems notions and concepts, most have in common a capacity to (a) actively mold program concepts to realistic constraints in the program environment, (b) take many plausible courses of action into consideration as part of the program planning process, and (c) seek program input and information from a wide variety of sources. Above all else these techniques "picture" programs as open systems that act in concert with the environment and which, in turn, receive input from it.

The single most definable quality linking these systems concepts is the emphasis and importance they place on the participation of a broad mixture of experts and laypeople who are capable of making judgments about the implications of programs and policies. Most of these techniques represent highly democratic processes in which individual opinion is heavily weighted. In contrast to other processes where program and policy alternatives must be weighed against their compatibility with some accepted ideology, systems techniques often represent democratically oriented approaches to the generation of alternatives and definitions of consequences, regardless of their relevance to any existing state of affairs. Thus, systems concepts tend to represent approaches to problem solving in which the criterion is least likely to be affected by any absolute ideology and is most likely to be defined by the best alternative available. This is accomplished by (a) providing for a maximum number of alternative courses of action to be discovered, many of which might not have been considered with less democratic methods, (b) encouraging a mix of opinions and viewpoints, often representing a consensus opinion reflecting the best features of individual viewpoints, leading to (c) identification of alternatives for which there is high probability that practical strategies, solutions, and methods actually exist.

We shall next turn our attention to some of the commonly cited criticisms of the systems approach. Our purposes for the following chapter are two-fold. First, some of the criticisms so lodged are quite valid; systems thinking as it has traditionally been sold in the marketplace has some notable limitations. Second, some of the criticisms of the systems approach have been unfounded and have masked some potentially good ideas.

This chapter has reviewed a broad array of techniques, procedures, and ideas that have been linked to the systems approach. From this chapter forward we will increasingly define our own view of systems. This view will share only some things in common with the systems notions and concepts presented in this chapter. Our belief is that some of the ideas reviewed here are potentially valuable to evaluative inquiry, whereas others familiarize readers with the breadth of systems notions and concepts. Our intent will be to borrow a few good ideas from the systems literature, meld these with some theories about values which were presented in Chapter 3, and integrate these with what we consider to be good ideas in the field of program planning and evaluation.

5

Systems Notions: An Evaluation Perspective

From the foregoing chapter on systems thinking and systems approaches one conclusion is obvious. The terms *systems thinking* and *systems approach* are not apt to mean the same thing to everyone. Because systems notions and concepts have evolved in a variety of fields, the semantic and conceptual confusion is not surprising, and it should be noted that the wide diversity in meanings of these terms has no doubt fanned the flames of skepticism and contempt for systems thinking. In this chapter we wish to provide a brief review and critique of these systems ideas from an evaluation perspective.

Barriers to the Use of Systems Ideas

Systems concepts have entered the realm of evaluation and have evoked responses ranging from open and accepting to hostile and rejecting, from cavalier and uninterested to very intense and personalized. A major obstacle to the application of systems concepts to evaluation has been a tendency, on the part of both advocates and critics, to misrepresent systems theory and methodologies. One form of misrepresentation has come from systems proponents who claim expertise in systems approaches despite their limited understanding of systems theory and limited experience with methodologies. The exaggerated claims of systems "salespeople" have produced an understandable skepticism and reluctance to deal with substantive issues in the application of systems concepts to evaluation. We turn now to a discussion of these barriers to the use of systems ideas in evaluation.

A TAXONOMY OF SYSTEMS PERSONALITIES

Hoetker (1972) provides an interesting and credible classification of "systems people." Speaking from the perspective of a school administrator, Hoetker identifies two types of systems people. Hoetker's first and smaller class represents true systems thinkers. These are individuals with a thorough understanding of systems and the motivation and judgment to apply that understanding, where appropriate. Hoetker's second class can aptly be called systems proponents. These are the "pushers" of systems approaches who may have something secondary to gain from their partisan systems stance. Hoetker subdivides this category into three groups.

The first group of systems proponents are the "fad followers," amateur enthusiasts who use the terminology of the systems literature and encourage the use and promulgation of some isolated systems approach of which they are aware. The fad followers' knowledge, however, is limited to rhetoric and they seldom can offer substantive ideas because of their limited systems expertise.

A second group of proponents are the systems consultants or contractors who make a living selling systems ideas. Hoetker generally assumes that consultants are genuinely skilled individuals with good intentions. They may or may not realize limitations on the power of their methodology in any particular settings, but they can be expected to work hard to provide the services for which they were contracted.

Hoetker's last subgroup of systems proponents are the "true believers." From reading Hoetker's words, one gets the feeling that he has dealt with a true believer a time or two for he waxes eloquently on the topic.

> The really bothersome systems type is the committed true believer. Whether in a university, a school, a government bureau, or a private firm, he is characteristically too narrow in his learning and too unreflective in his thinking to question seriously the latest Truth in which he has committed himself. . . . He is bothersome because he cannot be dealt with on a rational level as can the entrepreneur and, usually, the fad follower. One can only go along with the true believer or fight him . . . . If there is anything the schools do not need, it is more true believers, with whom they are presently well-overstocked and that in itself is reason enough to question the need for introducing yet another authoritarian strain of virulent certainty into an educational system that needs the right questions more than it needs another set of answers [1972, pp. 25–26].

The authors' experience with "systems people" of various sorts corroborates Hoetker's taxonomy. However, an unpleasant experience with a systems person or firm is not sufficient grounds to avoid everything labelled "systems." Alternatively one should not construe a positive systems contact as meaning that the systems approach will save the world. You may have encountered a competent systems consultant. Unfortunately, it may not always work out that way.

The personality issues brought to light by Hoetker (1972) parallel and compound the confusion surrounding systems approaches caused by a lack of uniform conceptual and pragmatic referents. It is the opinion of the authors that much of the criticism lodged against systems approaches can be traced to these two core problems.

SALESMANSHIP AND CHARLATANRY

The systems analysis business has been a highly profitable one for some individuals and the *marketing* of systems expertise has become a highly competitive arena in and of itself. Just like the lawyer, auto mechanic, plumber, and shop owner, systems analysts and thinkers need your business. There is considerable variability in how far these individuals will go in selling their systems services. Some have gone so far as to seriously damage the credibility of systems ideas. It is to this problem that we next turn our attention.

Hoos (1969) points out that, at that time, the market for system technology was booming.

> Wide spread and intense eagerness on the part of officials to "harness the powerful tools of technology" in the management of public affairs, in conjunction with a mixture of naiveté about the methods and political gamesmanship in high places, has brought about subsequent, expanding, and ever more lucrative contracts and grants for systems studies at local, state, and federal levels. The prospect of so bountiful a market has attracted prospectors of remarkable diversity in discipline, background, and prestige.

> There are aerospace and aviation firms, computer manufacturers and their multifarious subsidiaries, electronics companies, management consultants, accounting firms, and even public utility companies. Their ranks are swelled by university-based or -associated entrepreneurs, often in "institutes," from operations research, economics, engineering, and urban planning to mention a few. Additional contenders for contracts are the non-profit, albeit highly profitable, "think tanks," research institutions whose business it has become to solve public problems. Most of these are modeled after the RAND Corporation, originally the research arm of the U.S. Air Force. The newest, the Urban Institute, was established by the Johnson Administration to "analyze problems and provide teams of experts to solve them." Of its anticipated 10 million dollar annual budget, 80% will come from government sources [p. 18].

Hoos (1969) points out that contract proposals often provide the forum for salesmanship.

> The task of selection of the best contractor is complicated by the fact that propos-
> als for systems analysis of civil problems take the form of a mixture of jargon and
> salesmanship. Indeed, it is almost impossible to distinguish the merchandise from
> the merchandising, the evidence supporting the contractor's claim to competence
> to do the job from his generalized corporate "image." For example, all 50-odd
> proposals submitted in response to the State of California's call dwelt long on
> defense and aerospace contracts and subcontracts held, as though designing and
> building a rocket's launch system were proof positive of capability to design a
> system of criminal justice. Large sections of the proposals contain biographical
> material on all personnel of the contractor as well as curricula vitae of those
> individuals to be assigned to the project. But, as indicators of the company's
> ability, these are frequently irrelevant and generally unreliable. The company,
> research institute, or management consulting firm may have won an undisputed
> reputation in its field; it can truthfully count thousands of advanced degrees
> among its employees. These facts receive similar emphasis both in such an or-
> ganization's institutional advertisements and in its proposals. But these eulogies
> do not substantiate the claim to systems capability as needed in the social arena.
> Using this kind of justification ornithologists could compete for the same jobs as
> aerodynamics experts on the basis of experience with flying objects [pp. 41-42].

Our recent experiences with vendors of systems technology indicates that
the marketing practices described by Hoos are still in vogue 10 years later.
Particularly illuminating was a situation in which we were called upon to
independently evaluate the application of a contractor's specific systems
approach to a rather complex management problem in a military environment. Because this study had no precedents in terms of process, and even
the funding agency was unsure of what the study's results and potential
pay-offs might be, we were hard-pressed to elucidate some evaluative
criteria to use in making judgments about the study. We resolved this dilemma by isolating "claims" made by the contractor in the proposal. These
claims were then clustered into homogeneous categories which became the
basis for evaluative dimensions derived for the study. These claims took the
form of "this approach will reduce the time required to . . ." and "this approach will increase officers' understanding of . . . ." We decided that deriv-
ing evaluation criteria in this manner was ultimately fair to both the contractor (who made the claims) and the funding agency (who took the proposal as
a statement of services to be rendered).

Upon hearing of our intentions, the contracting firm became quite concerned about the "fairness" and "appropriateness" of our evaluative
criteria. We had naively assumed our strategy offered an opportunity to be
"fair" in an area where we had no expertise. The contractor explained to us
that these claims were admittedly extravagent "but this is how these proposals are usually written" and thus, they should not be taken that seriously. The
short of the story is that the president of the contracting firm consented to the
use of our criteria over the nervous objections of his subordinates and the

use of these criteria yielded both positive and negative conclusions about the specific systems analysis technique studied. This company president, an unusually perceptive individual, was quick to point out to his various department managers that in the future, "don't be surprised if people want to hold you to the claims made in a proposal." Our naive, yet common sense approach to this problem simply had not been encountered by this systems technology firm in a broad array of contracts in both the public and private sector and, quite simply, it surprised them.

Hoos's other comments about salesmanship are equally true. Considerable effort is devoted to the packaging of systems technology. Brochures are apt to be rather elaborate, expensive-looking productions that may unfold in unorthodox ways revealing seemingly simple concepts expressed in unique and complex ways. Lists of personnel read like a volume of *Who's Who*. The job titles of employees are similarly extravagant. Hoos's (1969) statement that "With the printing of business cards, individuals become 'Manager of Socio-Economic Systems' or 'Educational Systems Analyst' [p. 43]" reminds the junior author of his first day on the job doing some evaluation work for a systems technology vendor and being presented with a stack of business cards proclaiming him an "Associate Software Engineer." Apparently being just a plain vanilla evaluator wasn't good enough when interacting with clients. We must conclude that Hoos's comments about salesmanship and the systems approach are as timely today as they were then.

Systems approaches have been sold with a lot of "frill." In some cases this frill has served to obfuscate weaknesses (and strengths!) of a particular methodology. In other cases, the frill has alienated many potential users, particularly those insightful enough to see the frill for what it is. Thus, some good thinkers have "thrown the baby out with the bathwater" leaving other, perhaps less thorough, thinkers with the function of implementing systems tools, often in a less-than-ideal fashion.

THE PRACTICALITY OF THE SYSTEMS APPROACH

One of the foremost criticisms of the systems approach is that it is limited in practicality. This criticism most often arises as a result of some actual or perceived lack of applicability of some systems-oriented technique within some particular problem. However, it is often generalized to "the systems approach" as a whole. Perhaps overly optimistic claims were made by a contractor, or communication was less than perfect and a well-intentioned contractor's product fell short of his client's expectations. Beyond these problems of over-generalization and poor communication, however, there are some very real practical limitations on systems approaches generally.

Systems approaches have in common a certain thoroughness in their methodology. Generally, nothing is taken for granted or as given. These approaches usually insist on starting at Step 1 and proceeding carefully and methodically through some algorithm or heuristic process to arrive at an end point. While such a process can be essential to many highly complex problem contexts, such a methodology might "overpower" the problem in other settings. Methodological thoroughness can be an expensive, time consuming and labor-intensive enterprise. While some form of systems approach might offer a better ultimate solution, it may do so at too great a cost. Alternatively, the solution proffered may itself be impractical and not likely to be implemented.

Binner (1975), Coursey (1977), and Halpern (1977) point to the inapplicability of the systems approach to the evaluation of mental health programs. The following quote from Coursey (1977) is representative of these "practicability" criticisms.

> There are a number of problems to the systems analysis approach to evaluation. First, it is usually an expensive, complex process. In its more sophisticated form it requires greater measurement and more information data than are currently possible in any but the largest and most sophisticated mental health systems. Second, its theory or model of mental health is not yet adequately developed . . . . Third, how the evaluator is to compare the model and the real system is not yet worked out for mental health agencies, nor are system effectiveness indices well developed. Fourth, there are few illustrative studies of how this type of evaluation might be carried out (pp. 52–53).

Coursey is correct in pointing out that a systems analysis approach is apt to be "an expensive, complex process." He does not point out, however, that such an effort might be quite appropriate in very complex or traditionally intractable problem settings. In less complex contexts, some systems techniques would certainly "overpower" the problem, but not all systems tools are so thorough as to be impractical in these settings. Simply stated, there will always be a need to match the sophistication of a problem-solving technique to the breadth and depth of the problem being attacked.

Coursey's comment regarding measurement and information data requirements is also true, but is limited to systems approaches "in their most sophisticated form." We would rather conceptualize systems approaches, however, not in terms of their sophistication, but in terms of their scope and purpose. The specific scope and purpose of a systems analysis approach will dictate needs for information and specificity.

These criticisms by Coursey suggest that he sees the systems approach as some monsterously large, foreboding, and homogeneous methodology that will "swallow up" evaluative inquiry in the field of mental health. This "monster," however, exists only in fragments scattered throughout several disciplines and really poses no unified threat to anyone. Our fear is that these popular misconceptions are becoming second and third generation offspring. Coursey's (1977) second and third criticisms are cases in point. He states that its (presumably meaning the systems approach) "theory or model of mental health organizations is not yet adequately developed [p. 52]." We ask the rhetorical question "*Whose* theory or model?" Presumably, Coursey is assuming that some form of the systems approach has the responsibility for developing a theory or model of mental health organizations. It would seem that these tasks fall within the province of mental health professionals *themselves*. To doubt the usefulness of systems techniques to evaluation because some systems approach has not provided an overarching model for the organizational structure of mental health seems a bit presumptuous. Similarly presumptuous are his subsequent statements that there are no criteria for comparing "the model" and the "real system" or that this approach has not constructed acceptable indices of systems effectiveness.

Is Coursey implying that there are indices of *treatment* effectiveness in mental health and other social science settings, or of organizational efficiency or effectiveness in these settings? Are there adequate criteria for judging (evaluating) the effectiveness of these agencies and programs, or criteria for evaluating the evaluation of agencies and programs? Although inroads have been made in all these areas, including systems effectiveness indices, the answer is that effectiveness indices and criteria are not "well-developed" in any of these areas. To cite this lack in systems approaches to the evaluation of mental health programs is to obfuscate these more generic shortcomings in the field of mental health itself.

Coursey is most accurate, however, when he points out that there have been few reported applications of systems methods to mental health program planning and evaluations. Coursey points out that an adequate theory or model is *not yet* developed, that criteria for using systems models are *yet* to be developed, and that effectiveness criteria are not *as yet* developed. Thus there are few guidelines to follow.

THE VALIDITY OF SYSTEMS ASSUMPTIONS

Some writers in the field have questioned the practicality of systems approaches on the basis of their being too general and philosophically based. Hoos (1969) questions some of the assumptions underlying applications of systems theory concepts.

> The aerospace industry and others that have been closely associated with such large-scale enterprises as national defense and space efforts may claim a conspicuous degree of experience in the development of models of technically complex systems. It is unfortunate the word "system" can be used in conjunction with

nuclear weaponry and, say, elementary education, for it has led to the assumption and presumption that systems design, engineering and analysis as found in the first can be meaningfully and appropriately applied to the second, that they are somehow alike and amenable to the same treatment because they are "systems." Merely because, viewed in broad, general terms as an arrangement of component parts interacting in certain ways to achieve certain goals, any system fits this description, all systems become subject to the same "treatment." In the real world, there appears to be about as much justification for committing society's sundry malfunctioning systems to the care of a systems analyst whose sole claim to expertise is technical as to call a hydraulic engineer to cure an ailing heart because his speciality is pumping systems. Although the term "system" can be applied to both space hardware and social problems, the inputs are vastly different, as are the controls and objectives. In the engineered system, the components are tangible, the variables controlled, and the outputs identifiable. In the social sphere, the crucial elements often defy definition and control and do not behave according to a set of rules. There is no quality control of a social system; the test of its effectiveness is to a large extent a reflection of values and it is certainly not amenable to mathematical measurement [pp. 23-24].

Hoos makes a strong case for *not* pursuing a systems approach to social program planning and evaluation. Obviously, nuclear weapons and elementary education are very different and the same methodology cannot be applied equally to the two. However, Hoos conceptualizes the social sphere in systems terms by acknowledging the presence of inputs, controls, objectives, and systemic relationships between component parts in the system.

Hoos states that important elements of social programs defy definition and control. We ask, "Is this reason *not* to pursue some methodology that might bring some clarity to the area?" She asserts that critical components of the social sphere do not behave according to some set of rules. Considerable debate could be generated over this assertion. Obviously there is some order in our universe, even in the social sphere. We question not whether rules of order exist, but whether our traditional methods of inquiry will permit us or lead us to know these rules. We also take exception to the notion that a social system lacks quality control mechanisms. From our perspective such phenomena as social desirability, cultural mores, and public laws are active as general control mechanisms in most social systems.

Hoos is correct to point out the critical role played by values in social systems, particularly social service programs. She asserts, however, that values and programs are not amenable to mathematical measurement. Although there are developing methodologies to measure the benefits of programs (Halpern, 1977; Sorensen and Grove, 1978) and of values (Chapter 3 of this volume), these measurement methodologies are not yet fully developed. That a phenomenon cannot be easily measured mathematically does not imply that it cannot be the focus of inquiry. Even if mathematical measurement is impossible, most events lend themselves to qualitative description. The event's relation to other phenomena can be hypothesized, described in detail, and even tested empirically independent of mathematical measurement.

To be fair to Hoos, we should note that her criticisms cited here were proffered relatively early in the systems movement, when cutbacks in government spending on aerospace engineering brought an influx of unemployed "systems experts" into the arena of social program and policy planning. She draws a compelling picture of the "California Case" which she describes as an effort by the State of California to retain their well-educated scientists by redeploying their space-age technical skills to such social problems as overcrowded freeways, growing pollution problems, and soaring crime rates. It was reasoned that technology so powerful as to satisfy man's age-old curiosity about walking on the moon surely must have something to offer in terms of solving social problems.

Many of these early attempts were overly rigid, overly mathematical, and oversold techniques which fell short of expectations. Hoos (1969) reviews several specific contracts and describes the contractors' shortcomings. She does not acknowledge, however, that these were novel attempts to apply new and developing methodology to formidable, ever-growing problems. Some trial-and-error in such an undertaking is inevitable and is still occurring. Probably most upsetting to critics like Hoos is that "systems solutions" are pushed as though fail-proof, rather than as potentially useful. Whether systems approaches have something to offer is an empirical question. Failure on the part of systems proponents to acknowledge (or even recognize!) the equivocal nature of these early applications created some negative first impressions that have persisted.

Nor does it seem appropriate to criticize those early efforts on the basis of cost. While these were expensive studies that yielded marginal returns in some cases, most of these contracts were let at a time when the funding of social programs was less of an issue than it is presently. Massive funding for social programs was the order of the day, and when considered in context, the potential pay-off of these studies was judged as promising enough to justify funding. Most of these initial studies attempted to determine the feasibility of systems approaches to such problems as waste management, traffic control, crime in the streets, and the like, with the average contract amount being about $100,000. This average sum was no large amount, even in the mid and late 1960s. This is especially true when comparing these sums to the millions that were being spent for delivery of the services being studied.

Early criticisms, such as those of Hoos, have persisted in the literature until the present time. These perpetual criticisms lash out at a "systems approach" that no longer exists, having both evolved and decayed in the interim. Specific systems approaches, such as PPBS, PERT, and others reviewed in Chapter 4, have almost disappeared and simultaneously reappeared as various and sundry systems techniques and ideologies from several different disciplines. Many of these historic criticisms, therefore, simply are not as applicable as they once may have been. Yet they have found a place in the literature and are espoused almost as fervently today as a decade ago. It is as though many of the present critics are little aware of the intervening history of the decay, growth and development of systems concepts and technologies. Thus the "monster" we chided Coursey about earlier is in actuality only a ghost.

Halpern's (1977) criticism of the assumptions underlying systems analysis is exemplary of those critical of systems ideas. He presents a systems model, criticizes this model, and then uses these criticisms as a basis for justifying a cost-benefit model. The "systems" model Halpern presents is a simple Input-Process-Output model, similar to that presented in Figure 3.2 (p. 92). He proceeds to make the following points about the application of this model to mental health evaluation.

1. *Inputs.* Halpern states that the inputs to a human service delivery system are people and that it is safe to assume that no two input units will be the same because "no two people are exactly alike." He then states that "Such an assumption would clearly preclude the use of any systems model on purely logistical bases, and to the extent that the assumption is appropriate and/or correct, then the systems model is totally inappropriate [p. 78]." Having perused various systems methodologies and conceptual frameworks, we concur that an input-process-output model is a large part of some systems-oriented techniques, but we have encountered nothing in our experience that suggests that input "units" have to be precisely the same. While some linear programming models (Nagel and Neef, 1977), including some customized for program evaluation purposes (Bessent and Bessent, 1980), make this assumption, many others do not.[1]

2. *Process.* Halpern makes the case that because treatments are not often thoroughly specified for each client, "many sophisticated information systems (e.g., management information systems) are rather useless for program evaluation just because this fact is so often overlooked [p. 79]." We are aware of some systems analysis techniques that can be used to specify program components, such as LOGOS and SADT, presented briefly in the previous chapter. But it is unclear how lack of treatment specificity is a *unique* drawback of systems methodology. The work of Boruch and Gomez

---

[1]It should be noted that many statistical techniques assume that subject-clients (inputs in Halpern's model) are drawn randomly from the *same* population, but even here heterogeneity of variance is expected.

(1977), reviewed in Chapter 2, is one indication that this issue presents problems to traditional inquiry as well.

3. *Product.* Halpern makes the point that the outputs of human service programs are difficult to define. He uses this as a point of departure to explain a formula he and co-workers developed to attach monetary value to a client's response to treatment in mental health settings. While Halpern's efforts in this and other papers (Halpern and Binner, 1972) to develop some standardized measures of response to treatment are noteworthy, his criticisms of systems ideas based upon the difficulty of specifying outcomes cannot be traced to any deficiency in systems concepts themselves.

A similarly incomplete notion of the systems approach appears in House (1980). House equates systems analysis with (a) managerial science, (b) applied research, and (c) cost–benefit analysis. In the first case House suggests that the major audiences or reference groups for systems analysis are economists and managers. In the second case House describes the systems approach as one which

> defines a few output measures, such as test scores in education, and tries to relate differences in programs or policies to variations in the indicators. The data are quantitative, and the outcome measures are related to the programs via correlational analyses or other statistical techniques. Recently, experimental design has been more heavily employed and "planned variation" of program has been preferred to "natural" variation [p. 22].

In the final case, House sees the systems analyst almost totally consumed with questions of program cost versus efficiency: "Cost–benefit analysis of comparable programs is the hallmark of this approach [p. 25]."

From this tripartite definition of systems, the reader is left with a rather confusing array of concepts which may undergird the systems approach. But do they? We think House's description contains some stereotypes and artificial distinctions that warrant comment.

First, House's suggestion that systems ideas are directed primarily to (or are of most use to) managers and economists does not stand up in practice. Systems ideas and the systems approach has and can have many different audiences including managers, sponsors, constituents, program participants, *and the public.* Hoos's (1969) recounting of the diverse areas in which systems concepts have been applied, Stufflebeam's *et al.* (1971) provision of systems concepts in their evaluation model, and the many uses of systems concepts in government (Rivlin, 1971) attests to its broad appeal and use.[2]

---

[2]Readers may also note the newsletter of the American Educational Research Association Special Interest Group in Systems Research in Education for other audiences to whom systems studies have been directed.

As for managers being a major audience for the systems approach, we could not agree more. But, many evaluation approaches would claim managers to be among their "major" audiences. Managers could be expected to be no less of a major audience for the decision making and professional review (accreditation) approaches to evaluation which House also discusses. Of course, even within the same approach a major audience in one study can be a minor audience in another. In most evaluation studies the audiences to whom the evaluation is directed depends upon many factors, not the least of which are the political forces at work (Brickell, 1976), the negotiation and accommodation that has taken place within the policy-shaping community (Cronbach et al., 1980), and the extent to which multi-partisan decision making has occurred (Lindblom and Cohen, 1979). Thus, we do not find the concept of *audience* to be a useful way of distinquishing among evaluation approaches.

Second, House's discussion comes close to equating the systems approach to that of applied research. House cites the use of test scores, pre- and post-testing, data aggregation, and the use of normal-curve equivalents as other characteristics of the systems approach. But, these concepts and procedures are characteristic of quantitative inquiry in general and should not be construed to represent the totality of the systems approach. To the contrary, van Gigch (1978) points out that, instead of relying exclusively on analysis and deduction, the systems analyst needs to synthesize and to be inductive. Instead of relying strictly on formalized methods of thinking he or she must take into account:

1. Informal reasoning processes such as judgment and intuition
2. The weight of evidence stemming from few observations and small chance of replication
3. Predictions based on weaker evidence than explanations
4. More domain discontinuities and the importance of the unique event (p. 198).

With regard to House's final assertion that cost-benefit analysis is the hallmark of the systems approach, we note that each of three leading texts on the systems approach devote only the equivalent of a single chapter to this topic. (See Churchman, 1968; Emery, 1978; and van Gigch, 1978.)

Before leaving this topic the authors would like to point out that our specific criticisms of Coursey, Halpern, Hoos, and House should not be taken as representative of our sentiments about these writer's contributions to the field of program evaluation in general. Each has made considerable contributions to the development of evaluation theory and practice. We do, however, take exception to many of the criticisms leveled at systems approaches which have conveyed limited and sometimes distorted notions of systems concepts and systems thinking.

THE EFFICIENCY OF SYSTEMS APPROACHES

Another limitation often cited in discussions on systems approaches is that they tend to substitute form for content. Of the criticisms we have reviewed thus far, this one is particularly salient. Many systems approaches demand a level of throughness and documentation not part-and-parcel of more traditional methodology. The following quote from Senator William Proxmire (1970) is illustrative of this criticism.

> A great many gimmicks that use manpower and create vast paperwork should be done away with. The services should reduce or stop, because of their obvious ineffectiveness, such programs as PERT, PEP, Value Engineering and the Zero Defects concept. The mere list of names makes one shudder. We should resist new gimmicks, such as Total Package Procurement and the Milestone Buying System. All of these essentially substitute form for substance. A saving of $500 million should be possible from cutting these nonessential items alone [p. 236].

We take the position here that it may not be the techniques per se that are responsible for the wastes such as those reported by Senator Proxmire. Although some of these "time savers" and "advanced organizers" are no doubt wasteful, others may be potentially helpful in program planning, development and evaluation. It might be that some of this waste is not the result of the techniques per se but rather how well the techniques are matched to the requirements of the problem context to which they are being applied. The failure of this type of match usually results in the all too familiar case of overpowering a problem with an overly sophisticated solution

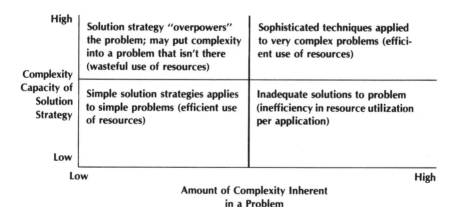

Figure 5.1. Conceptual space of two complexity continua: problem complexity and complexity capacity of solution strategies.

strategy and leads to the kind of criticisms discussed earlier in this chapter. Optimally, the solution strategy's capacity for dealing with complexity should match the amount of complexity inherent in the problem. These two ideas are expressed in Figure 5.1.

Current Systems Notions in Evaluation

The preceding pages of this chapter have focused on impressions and criticisms of various systems notions and concepts that, in the main, have been borrowed from other disciplines and applied to program planning and evaluation problems. In addition to borrowing systems ideas, the field of evaluation has developed some systems-oriented notions and perspectives of its own. It is to the evolution of systems ideas within the field of evaluation that we now turn our attention.

The systems notions specific to evaluation can be divided into three related categories. Each represents ways of dealing with or thinking about the complexity inherent in programs. First, the field of evaluation has responded to increasing complexity in the information needs of different actors and interests surrounding any program effort. Second, evaluation has made some attempt to admit to the organizational realities that constitute basic components of virtually any program. Third, the field of evaluation as a whole has admitted to some serious limitations in its traditional methodology and is striving to develop alternatives. In all three instances, the response by the field has at its roots been a systems-oriented one.

COMPLEXITY IN INFORMATION NEEDS

Suchman's (1967) view of programs as systems is representative of early systems-oriented ideas found in the evaluation literature. Like many of the early writers in the field, Suchman viewed systems as larger-order efforts to change aspects of society (e.g., health systems, education systems, welfare systems, and the like). These large-scale activities comprised narrow-scope organizations and programs. Evaluation, within this context, usually proceeded without addressing these different perspectives on a given program. Thus an evaluation that met expectations at the program level may have been totally inappropriate or irrelevant at the system level. Of course, the opposite may also be true.

Suchman espoused a dichotomy between *system* evaluations, which would focus on broad-aim objectives, policy decisions, and adjustive recommendations, and *program* evaluations, which are attempts to relate specific program activities to specific program outcomes. In this manner the system and program concepts were juxtaposed but the systemic properties of programs themselves were not elucidated. That is, programs were seen as component parts of a larger system, but they were not understood as possessing systemic properties in and of themselves.

This view of programs and systems is also salient in the work of the Phi Delta Kappa National Study Committee on Evaluation (Stufflebeam *et al.*, 1971).

> In education, there is some agreement that the term *system* is a general term which subsumes specialized efforts called programs. For example, a state system subsumes a social studies program, a mathematics program, a science program, a language arts program, etc. Evaluation of programs is carried out in the context of the state educational system. The output of the city system serves as input for other systems, both within and outside the state, and may also be thought of as input by decision makers in other systems and at the national level [p. 130].

Stufflebeam and his co-workers, however, were more sensitive to the evaluator's dilemma of having to provide evaluation information for system decisions *and* program decisions. At that time there was little help for the evaluator in the field who was feeling the sharp edge of both sides of the razor. The work by Stufflebeam and his colleagues on the PDK Study Committee attempted to fill this void by presenting a well-conceptualized systems view of evaluation that combined the committee's considerable evaluation experience with their knowledge of traditional as well as developing technology in the field, particularly the use of computers and management information systems.

These authors advocated a multilevel evaluation model that acknowledged both qualitative and quantitative differences in information needs at various levels of decision making. Some decisions were concerned with individual programs whereas others were focused on more generic system performance levels, as had been suggested by Suchman. The work of Stufflebeam and his colleagues constituted one of the most thorough as well as earliest elucidations of a systems perspective in evaluation. Although their systems ideas are somewhat dated, we shall have occasion to reference several of them below.

Particularly notable of the systems ideas presented by Stufflebeam *et al.* (1971) was that the input–process–output conception of a program is an oversimplified view of what happens in programs. They replaced the "empty" conceptualization presented in Chapter 3 (Figure 3.2) with a much richer conceptualization acknowledging the systems properties of programs. This conceptualization is presented in Figure 5.2. This figure can be viewed as a graphic depiction of the CIPP evaluation model (*Context, Input, Process, Product*).

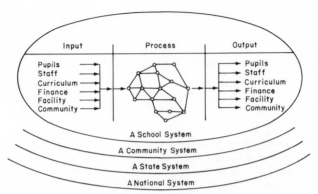

Figure 5.2. **Stufflebeam *et al.*'s (1971) system conception of the CIPP evaluation model.**

The CIPP model provided a rationale for viewing programs first in terms of the program's context. The PDK Committee suggested that context evaluation be pursued in two modes.

In the contingency mode, context evaluation searches for opportunities and pressures outside of the immediate system to promote improvement within it. This involves exploratory probing within loosely defined boundaries—study visits to other systems; exploration of the research and development literature; outside consultants; brain-storming retreats; assessments of community values, attitudes and priorities; and so forth. This mode of context evaluation also probes the future. Societal needs and values, technological advances, political trends, economic developments, population statistics, etc., are projected [p. 219].

In the congruence mode of contextual evaluation, indices of actual and intended system performance are compared. Federal, state, and local laws and policies are operationalized into operating procedures, line-staff and administrative organizational schemes, budgets, calendars, salary schedules, planning activities, interagency contacts, and the like. The congruence mode of contextual evaluation is concerned with monitoring the system to determine whether these operationalized goals and schedules of events are being achieved as intended or required.

The contingency mode of evaluation focuses on more dynamic aspects of the environment than does the congruence mode, and together they serve to describe in detail the constraints on the system. Context evaluation, as Stufflebeam *et al.* (1971) define the term, serves to specify the environmental presses that will moderate or constrain what some programmatic activity might have as its goals, what form it might take, what clientele it may serve, and what might be its ultimate purpose.

The other three types of evaluation described by the PDK Committee (input, process, and product evaluation) were more familiar to evaluation and other program personnel, but even these were discussed within a novel conceptual context for most readers. In particular, input evaluation (determining the optimal utilization of resources to meet program goals) represented a new wrinkle in that input evaluation had traditionally been viewed as a management (as opposed to an evaluation) function. All three of these types of evaluation were discussed within the context of a tripartite model of planned change. The model provided some means for evaluators to judge when to do what in a variety of evaluation contexts and the evaluation model itself was offered within the context of information dissemination and utilization in decision making.

However, the well-conceptualized systems view of evaluation presented by Stufflebeam and his colleagues on the Phi Delta Kappa Committee (Stufflebeam et al., 1971) fell on deaf ears. The failure of this group to strike a responsive chord among practicing evaluators is due to several factors, not the least of which was the emphasis these writers placed on decision making vis-à-vis evaluation. Because of this emphasis on decision making, much of the conceptual richness of the PDK volume was lost on its readership (see Chapter 2 of this volume for a discussion of this issue). What most readers took away with them from that book was not a systems perspective on evaluation in the generic sense suggested by Figure 5.2, or even a conceptual or philosophical "feel" for the development of multifunction information systems, but instead a view of evaluation as the handmaiden of decision making, or a justification of the evaluator's role to collect and pass on information.

It is our belief that the world of evaluation was not ready for the level of exposition offered by this committee. Evaluation was a neophyte as a profession, not ready or capable of responding to these forward ideas. Individual evaluators of the time were undergoing minor existential crises as they were bombarded with opinions and evidence of their failure to depict "truth" to every conceivable interest group. Furthermore, it is usually acknowledged that the level of "technical expertise" of many of the early evaluators (or at least their evaluations) was not very high. The field of evaluation was at the time in need of simplification. In addition, the PDK Committee itself was in uncharted territory, attempting to elucidate a way of going about evaluation that would somehow ease evaluation's "ills." Thus they were attempting to tell others how to do something that had not been done before. There was no consensus on the role of evaluation and little definitive knowledge on how to design information systems for program monitoring and evaluation.

Therefore, although the evaluation concepts advocated by the PDK Committee represented a complexification of the task of evaluation, they were interpreted as a simplification of the role of evaluation. What emerged from this classic volume was the definitional idea that evaluation exists to serve decision makers. The work of this group simply did not foster a systems *view* of programs nor did it stimulate readers to pursue that vision on their own. While they introduced a rich variety of systems ideas into the evaluation literature, a unified systems view of programs did not emerge. Nor has it emerged in the field of evaluation to date.

Thus the CIPP model this group proposed was little understood; it was more often taken as a step-by-step algorithm than as the conceptual heuristic for thinking through an evaluation that it was. Moreover, the PDK Committee's suggestions placed new demands on evaluation and evaluators. No one likes to have their shortcomings pointed out and some evaluators and research methodologists may have resented statements such as the following:

> Perhaps the greatest challenge of the many confronting evaluators at this point is to overcome the idea that the methodology of evaluation is identical to the methodology of research [Stufflebeam *et al.*, p. 140].

> The construction of a new general theory for collecting, organizing and analyzing data will be a formidable task. The analysis above seems to indicate that it is possible but provides little insight into how it should be done. The careful attention of statisticians is clearly needed.

> Even if this general task were already accomplished, many other methodological tasks remain. Among the most major of these is the development of instrumentation appropriate to the evaluation tasks outlined here. We can no longer be cavalier about dismissing as intangible variables that are not evidently measurable by paper-and-pencil techniques. The requirements of process evaluation in particular, and other kinds of evaluation as well, require that we devise unobtrusive techniques that can be used in real-world situations without diverting attention. We need to devise approaches that permit the utilization of information such as testimony that we have heretofore felt was too "subjective" to be useful. Since the courts have operated for centuries largely on the basis of testimony, why should social scientists not be able to devise ways of handling testimony in evaluation settings? Until we solve these and similar problems that put tools into the hands of the evaluator, we shall have to be content with mediocre evaluations. Methodologies will also have to be devised in the area of design. The design problem tends to overlap the two roles of the evaluator—the technical and interface roles—but we have chosen to regard this problem as primarily technical in nature . . . [Stufflebeam *et al.*, 1971, pp. 338–339].

We should point out that while the ideas presented by Stufflebeam *et al.* (1971) may not have had their intended impact, these individuals did have considerable influence on the field of evaluation through their training programs which inculcated new evaluators with these ideas. That influence should not be discounted.

Some writers in the field of evaluation, such as Halpern (1977), have interpreted a systems perspective only as a simple Input–Process–Output model. It is clear from Figure 5.2, however, that one's systems perspective can be considerably more encompassing and thorough. Even Figure 5.2 can be considered a rather limited systems view of a program. The model could be enriched to indicate the change experienced by some program part (students or community, for example) as a result of experiencing a program (the "process" in Figure 5.2). In addition, the logical flow between specific program activities could be depicted in detail in the process component of Figure 5.2. Furthermore, the model could attempt to elucidate the interrelations between the program and the larger systems of which it is a part as well as to describe the interface between these successively higher order systems.

The field was more responsive to the message offered by this group with regard to the computerization of evaluation. Developing computer technology coupled with the basically quantitative, computer-oriented research background of evaluators gave rise to the development and elaboration of management information systems in human service and educational programs. Information system planning and implementation has become a highly technical evaluation subspecialty that addresses those concerns aired by Suchman, Stufflebeam and his cohorts, and other writers of the day.[3] Simply stated, those concerns were "*How* do you get *what* information to *whom?*" The work of Stufflebeam's group was seminal in this development.

ORGANIZATIONAL REALITIES

Related to the issue of increased complexity in information needs about programs, a tangent of the evaluation literature has addressed the organizational realities surrounding a program. Stufflebeam *et al.*'s conceptualization of evaluation, particularly context evaluation, reflects this shift toward a consideration of the environmental context in which a program operates. This movement was opposite the tendency in evaluation to use experimental designs that attempt to control, hold constant, or otherwise dismiss contextual considerations bearing on the functioning of a program. In more recent times the organizational context of evaluation has become an increasingly popular topic, with considerable emphasis now being given to the place of evaluation in an agency's organizational chart, to the dilemmas of external as opposed to internal evaluation, and to the juxtaposition of evaluation practice vis-à-vis management functions.

[3]For the reader interested in current developments in the use of management information systems in human service programs the following references are suggested: Sorensen, J. E. and J. R. Elpers. Developing information systems for human service organizations; Broskowski *et al.* A management information and planning system for indirect services; Elpers, J. R. and R. L. Chapman. Basis of the information system design and implementation process. All in Attkisson *et al.* (Eds.), *Evaluation of Human Service Programs.* New York: Academic Press, 1978.

Etzioni (1960) provided one of the earliest "context" perspectives on evaluation. He argued that one needs to view programs as open social systems with a myriad of formal and informal goals and functions. He argued that goal attainment models foster a narrow evaluation focus that precludes understanding the organizational realities surrounding a program, realities that are often the primary determinants of what actually happens in a program.

Organizations, agencies, programs, and individuals engage in a number of activities besides maximizing attainment of a goal or subset of goals. Among these are survival of the agency and protecting one's own job, maintaining positive relations between staff members, and bringing about positive relationships between the program and clients, the community and other agencies. Etzioni emphasized the interrelations within the organization and the interface between the agency and its environment. He insisted that an evaluation of an agency or program should examine these issues as well as assess whether or not a program achieved particular goals. Etzioni argued that an evaluation that emphasized the accomplishment of some specific, usually narrow goal to the exclusion of these prepotent environmental presses and processes yields evaluation results of questionable validity.

Whereas the view of Stufflebeam *et al.* (1971) and Suchman (1967) emphasized the role of programs as component parts of larger systems, Etzioni (1960) clearly viewed programs as systems in and of themselves. It should be noted that Etzioni's work, like that of Stufflebeam's PDK Committee, was only marginally acknowledged by the field of evaluation at the time.

Thus, when a program administrator informed an evaluator as to the goals of a program, the evaluator scurried off on a mission to determine if those goals had been met without even once considering the appropriateness of those goals. Etzioni's view would raise the following questions: "Did the administrator choose easily achieved goals which he or she knew the program had met, thus ensuring that he or she would appear to be running a 'tight ship' and generally doing a good job, which might then constitute a means or justification for continued program funding, a personal pay raise, or promotion?" Or, "Were the goals selected because the administrator knew that these were highly valued by some staff members and their selection constituted a means of promoting staff cohesiveness and loyalty, although the goal selected in actuality only minimally indicated program impact?"

These are illustrative of the kinds of questions Etzioni's approach would suggest. Similar questions could be posed with regard to all aspects of program functioning and program decisions, not only the goal selection process. However, the world of evaluation was not ready for the kinds of questions suggested by Etzioni's open social system view of programs. Perhaps owing to evaluation's fledgling status and organizationally subservient role, it may have been professionally suicidal to pose such questions (in some settings it still is). For whatever reason(s), few evaluators tried to take Etzioni's ideas and apply them in evaluation practice.

A notable exception was the work by Schulberg and Baker (1968) which echoed and elaborated Etzioni's concerns about the goal-attainment model. Whereas Etzioni addressed organizational effectiveness, Schulberg and Baker juxtaposed these ideas vis-à-vis the evaluation of public health programs. Borrowing from Etzioni, they proposed a "system model" of evaluation that seeks to study a multifunctional unit that, in addition to achievement of goals and subgoals, is concerned with "the effective coordination of organizational subunits; the acquisition and maintenance of necessary resources; and the adaptation of the organization to its own internal demands [p. 1252]." Rather than focus on a specific objective, their system model attempts to understand goal-attainment within the context in which the program operates. In this manner, program outcomes may be understood only within the context of program inputs and constraints.

Wildavsky (1972) has provided a classic analysis of evaluation vis-à-vis organizational issues. He points out the seemingly inherent conflicts between the needs of a program and its staff, and the willingness to evaluate and change program policies and procedures. He describes a utopian model of a self-evaluating organization, warning that general personality considerations or foibles of staff will significantly moderate or constrain the degree to which the organization can evaluate itself.

Wildavsky's (1972) article offers an insightful systems-oriented view of the organizational context in which evaluation must ultimately operate. The major contribution of this work lies in its articulation of the very practical organizational constraints that are incompatible with the process of evaluating and changing the organization. Although the article has become a standard reference (particularly with regard to internal versus external evaluation issues), it has had little impact on the actual conduct of evaluation in practice. Wildavsky poignantly pointed out some problems about the context in which evaluation occurs; unfortunately, the problems he elucidated are not easily dealt with. Subsequently, his systems-oriented notions of the organizational context of evaluation did not stimulate much improvement in evaluation practice.

It should be noted that Wildavsky is not the only writer to address this issue and make suggestions for improvement. Much of the recent literature on utilization of evaluation results addresses organizational realities surrounding the evaluation of programs. Representative of these works are Attkisson, Brown, and Hargreaves (1978); Broskowski and Driscoll (1978); Bunker (1978); Patton (1978); and Weiss (1977a). Murrell (1977) makes the point that consumers, the service organization and evaluators should be included in the evaluation as three equal partners. He argues that these three parties have agendas that may conflict and that collaboration is essential to resolving such conflicts. Like Etzioni and Schulberg and Baker, he views programs as open social systems. These systems determine the form and process of the program implemented.

> The typical first step in many program evaluations is to ask the agency to describe its goals . . . . Yet only rarely does "continue to exist" appear on such a list. The evaluator can himself safely supply "survival" as the first priority goal and must recognize that much of the information he collects will be constrained by this goal. It seems to me that the most difficult problems of conducting program evaluations stem from organizational realities rather than from lack of knowledge of methods and designs. It seems essential, therefore, to fully recognize the real multi-functional complexities of these open social systems [pp. 259-260].

The idea that our programs and the administrative structures that manage them have a life of their own has recently evolved in the evaluation literature. Weiss (1972c, 1977a) continues to point out the inherent political nature of evaluation. Horst, Nay, Scanlon, and Wholey (1974) illustrate the "management vacuum" existing in many of our federally funded programs. In Chapter 2 we discussed the lack of continuity in policies, goals, and personnel that result from our changing executive administrations every 4 or 8 years. Horst et al. (1974) make the case that such contextual considerations render many programs "not evaluable."

Friedman and Friedman (1980) take the ideas of organizational realities vis-à-vis evaluation functions to their furthest extremes. They define the roles of federal employees first as a labor union and second as a lobby, both of which quite effectively protect their members' self-interests. When considering the picture the Friedmans paint of the federal bureaucracy, one must agree with Etzioni's (1960) original argument that indeed there may be much that occurs in the organization that is related only tangentially, or not at all, to the stated goals of a particular program.

Our point is that much has been said about the organizational context of evaluation and that these commentaries represent a systems view of the work of evaluation. Although numerous suggestions have been offered about the linkage of evaluation to decision making, little has been suggested about how to examine the systemic contextual considerations that can moderate greatly the kind of evaluation that goes on. Again, we have some literature that implicitly and explicitly proposes a systems *perspective* on evaluation, but little effort has been devoted to using these organizational insights in evaluation practice.

EMPHASIS ON QUALITATIVE METHODOLOGY

Recently, considerable interest has emerged in qualitative research methodology in a number of fields, including program evaluation. Much of this work explicitly or implicitly embodies a systems perspective. In Chapter 3 of this volume we discussed the work of Guba on naturalistic inquiry. The general theme of that chapter was that values were of a higher order than are program processes or outcomes themselves and that evaluators needed to be "value-oriented" if they are to know if programs are actually doing the things we want them to do. Guba distinguished several models of qualitative data gathering and analysis that encompassed this systems view of values and programs.

Within the field of evaluation, we have witnessed a virtual explosion of books and articles devoted to the development and/or use of qualitative methodologies for program evaluation purposes. Short papers by Weiss and Rein (1972) and Parlett and Hamilton (1976) are illustrations of these. The general themes of these works are displeasure with the status quo of traditional evaluation methodology, which is almost exclusively quantitative, and a call for qualitative alternatives.

The primary reason that we classify this qualitative methodology literature as both value- and systems-oriented is based on the kinds of questions that are asked and the way data are gathered within the methodological orientation. Rather than simply document program success/failure in terms of outcome "scores" of some sort, qualitative methods yield richer data on which subsequent decisions can be made. Weiss and Rein (1972) state that

> the need in the study of broad-aim programs is for a conscientious attempt to find the forces which shaped the specification of the program, the nature of the opposition it encountered, the reasons for such failure as occurred, [and] the program's unanticipated consequences. Then, in addition the research might identify the anticipated changes which occurred and the ones which did not. The issue in evaluation of broad-aim programs is not "Does it work?" but rather "When such a program is introduced, what then happens?" [p. 240].

The evaluation needs identified by Weiss and Rein (1972) are value-oriented because their answers require an understanding of the means–end continuum. They are systems-oriented because their answers require one to consider the larger program, or system of which the program is a component part. Clearly, traditional, numerically based analyses are not going to address all of these value- and systems-oriented questions in a cost-effective way.

The broader scope of qualitative methods makes them better matched to these systems- and value-oriented kinds of questions. Weiss and Rein (1972) make the following comments in their comparison of qualitative analysis with the more traditional approach.

> Evaluation is the appraisal of the extent to which a program realizes certain goals. When evaluation is performed by experimental design, the goals in terms of which the performance of the program is evaluated are those imbedded in the program at its inception. The value framework used for appraisal is for the most part that of the program's administrators, and though there is some small opportunity to do so, there is no incentive, in the experimental approach, to use alternative sets of values to develop criteria.
>
> Qualitative study presents quite another situation. Now it is possible to describe the extent to which the program realized its initial objectives, but it is also possible to appraise the extent to which the program realized other goals as well. The investigator can ask whether members of the target population have suffered losses—perhaps by attack on a coherent way of life—as well as gains. He need not restrict his attention to the target population, but can describe what seem to have been the consequences of the program for individuals in other sectors of the community. The investigation can consider the program not only from the perspective of clients, but also from the perspective of other agencies with which it may have been competitive, or of local government, whose managers it may have by-passed. If he wishes, the investigator may evaluate the program from a radical perspective and consider the extent to which the program has patched up a destructive system rather than initiating fundamental changes.
>
> Qualitative appraisal permits values to re-enter evaluation openly, instead of only implicitly as when the experimenter limits his attention to whether the program works. In addition, it permits the development of a set of appraisals, each from a different perspective. These are by no means the least weighty recommendations for this approach to evaluation [pp. 247–248].

In a similar vein Patton (1980) elucidates the relationship between qualitative research methods and a systems view of programs.

> Researchers using qualitative methods strive to understand phenomena and situations as a whole; evaluators using qualitative methods attempt to understand programs as wholes. The researcher strives to understand the gestalt, the totality, and the unifying nature of particular settings. This holistic approach assumes that the whole is greater than the sum of its parts; it also assumes that a description and understanding of a program's context is essential for understanding the program. Thus, it is insufficient simply to study and measure the parts of a situation by gathering data about isolated variables, scales, or dimensions. In contrast to experimental designs which manipulate and measure the relationships among a few carefully selected and narrowly defined variables, the holistic approach to re-

search design is open to gathering data on any number of aspects of the setting under study in order to put together a complete picture of the social dynamics of a particular situation or program. This means that at the time of data collection each case, event, or setting being studied is treated as a unique entity with its own particular meaning and constellation of relationships emerging from and related to the context within which it exists [p. 40].

The holistic perspective attributed to qualitative research methods by Patton is similar to our own systems view of programs. The burgeoning field of qualitative research methods has indeed taken on the proportions of an entire movement. (See Rist, 1980, for a whimsical view of these developments by one of the pioneers in the field.) Considerable debate has been waged on the topic. For our present purposes we wish only to show that systems-oriented notions of evaluation have been kicking around in the literature under the rubric of qualitative research methods for some time. For readers interested in pursuing this topic the following references are suggested: Britan (1978); Campbell (1978, 1979a,b); Filstead (1979); Guba and Lincoln (1981), Patton (1980); and Reichardt and Cook (1979).

Summary

In this and the preceding chapter we have tried to illuminate the wide diversity in theory, methodology, and practice that has come to be lumped together under the commonly heard terms *systems approach, systems thinking,* and *systems theory.* We have reviewed common criticisms found in the literature of several different disciplines and attempted to interpret the evolution of systems ideas, particularly in the last 10 years. Because we intend to make a case for a systems-oriented view of program evaluation, it seems timely to examine in some detail what has come before. How can we untangle this web of diverse theory and procedure, of overzealous advocacy, and of scathing, cynical fault-finding that obscures many legitimate attempts to deal with complex and difficult problems?

Five points should be salient from the foregoing review. First, systems thinking is not a static, unitary concept as, for example, are the principles of operant conditioning. Systems ideas are both varied and evolving.

Second, many of the claims *and* criticisms of systems notions as applied to social programs *are not* relevant in considering the potential fruitfulness of systems thinking for program evaluation. In the literature some claims and/or criticisms specific to a particular systems notion have been erroneously generalized to others. Other claims/criticisms have become outdated and are no longer applicable to systems concepts and notions. Yet many of these claims and criticisms have become reified in the literature, though the truths they may have accurately portrayed at one time no longer hold. They would be harmless except for the fact that these criticisms keep some systems notions suppressed. If not "aired out" intellectually and commonsensically, the potential heuristic value of these ideas will go unmined. Alternatively, excessive claims have brought about unrealistic expectations to the point that disappointment in and distrust of systems notions abound in many professions, including evaluation.

Third, the above application of systems ideas to social, political, and other "nonphysical" problems is a developmental process. We can expect successively finer and more appropriate (useful) applications as we better discern the parameters constraining this applicability. This development has been carried as far as it can be by systems experts. The responsibility is now on subject matter experts to apply these notions within their respective disciplines. If these systems notions have anything to offer the related worlds of program planning, management, and evaluation, it will be up to those individuals to make it happen. This is where the field is developmentally.

Fourth, there is a common element running through the quite varied notions and procedures reviewed in this and the preceding chapter. This common thread is a viewpoint or philosophy in which one is aware of, or at least does not ignore, the systemic relatedness in the world (i.e., that everything is related to everything else). It is the view that our world is richly complex and interdependent. Implicit in this richness is that we cannot fully comprehend what goes on in the world if we try to understand it in a fragmented, one-piece-at-a-time fashion. There is just so much more going on at any one time than we can grasp intellectually.

Fifth, systems-oriented notions and ideas have been around for some time. Although many of these ideas are systems-oriented, a systems perspective on evaluation has not been elucidated. These notions and concepts represented pieces, but not the whole picture.

We would like to add that absent from the literature of program evaluation has been the notion of General Systems Theory. This is somewhat surprising as the early writings of the General Systems theorists (Bertalanffy, Boulding, Buckley, Churchman, Etzioni, Laszlo, among others) appear to be at the root of much of the emphasis being placed today on the context, setting, or environment within which events take place. While we do not claim that General Systems Theory and more recent movements in the field of program evaluation (such as ecological psychology, ethnography, naturalistic inquiry, and qualitative methods) are isomorphic, we are suggesting that a cogent theoretical and conceptual rationale for these movements lies within General Systems Theory. In the following chapter we will discuss the nature of General Systems Theory, which indicates the need for and logic of these movements as well as provides conceptual and theoretical rationale for the reform of program evaluation.

The capacity for complexity in our program planning and evaluation methodology has not matched that existing in the world we are to evaluate. There is some consensus that we are in need of some heuristics to help us think about programs in a more efficacious manner. Calls for broader conceptual and methodological perspectives in program evaluation have been voiced by many of the most respected of program evaluation theorists and methodologists in the last 30 years (Guba, 1969; Koch, 1969; Mann, 1972; Riecken, 1953; Schulberg and Baker, 1968; Scriven, 1967; Stufflebeam *et al.*, 1971; Suchman, 1972; Thompson, 1967; Weiss, 1972b,c; Weiss and Rein, 1972). More recently, well-respected traditional methodologists such as Robert Boruch (Boruch and Gomez, 1977), Donald T. Campbell (1978, 1979a,b), Gene V. Glass (1979), and Lee Cronbach (Cronbach *et al.*, 1980) revealed trends in their thinking toward broader conceptual and methodological bases for thinking about programs. Conceptually, the recent thrust has been toward a conceptualization of programs as systems (Attkisson and Broskowski, 1978; Broskowski and Driscoll, 1978; Horst, Nay, Scanlon, and Wholey, 1974; Murrell, 1977). Methodologically, the thrust has been toward qualitative research methodology (Campbell, 1978; Cook and Reichardt, 1979; Patton, 1980), particularly anthropological and ethnographic studies of programs (Bogdan and Taylor, 1975; Britan, 1978; Glass, 1979; Guba and Lincoln, 1981; and Parlett and Hamilton, 1976).

Thus the idea that the philosophy and methodology of evaluation needs to advance is neither new nor forgotten. It is our contention that the direction of movement called for by many prominent evaluators in the field is consistent with the general systems notions and ideas presented in this and the previous chapter. We believe that adopting a systems view of the world and utilizing methodology consistent with that view will make us better program planners, managers, and evaluators and that systems notions can serve the valuable role of heuristics in thinking about our programs. It is to this end that we devote the remainder of this book.

Well, then, what is the systems approach? On the one hand, we must recognize it to be one of the most critical problems we face today, the understanding of the systems in which we live. On the other hand, however, we must admit that the problem—the appropriate approach to systems—is not solved, but this is a very mild way of putting the matter . . . . It's not as though we can expect the next year or a decade from now someone will find the correct systems approach and all deception will disappear. This, in my opinion, is not in the nature of systems. What is in the nature of systems is a continuing perception and deception, a continuing reviewing of the world, of the whole system, and of its components. The essence of the systems approach, therefore, is confusion as well as enlightenment. The two are inseparable aspects of human living. [Excerpted from *The Systems Approach*, by C. West Churchman, copyright © 1968 by C. West Churchman, pp. 230-232, reprinted by permission of Delacorte Press.]

6

Programs and Systems

In the last two chapters we have presented both a review and a critique of systems approaches. We have pointed out the varied forms this type of inquiry has taken and have reviewed the problems and difficulties that have arisen from these developments. If these chapters left the reader with the impression that "systems," as a specific discipline or technology, is rather muddled and messy, then we accomplished a large part of our purpose.

In this chapter we hope to show that despite the muddle, misunderstanding, and malfeasance, there are some rich concepts and ideas in this literature that can be useful to program evaluation. There are some common features in these diverse systems sources, both conceptual and practical, that may address some of program evaluation's inadequacies and weaknesses. These weaknesses include the lack of a guiding theory behind the developing field of evaluation and the fact that our traditional evaluation designs, which have utilized experimental design principles almost exclusively, do not always provide adequate or meaningful measures of program effects.

In this chapter we propose that a systems-oriented perspective can serve to alleviate some of evaluation's ills. This perspective is not an original one; it borrows material from very diverse fields of inquiry. A synthesis of these ideas as they are applied to program evaluation seem particularly timely for two reasons. First, we appear to be at a stage of increasing dissatisfaction with traditional methodological paradigms, a view suggested by Kuhn (1970). This dissatisfaction is being felt most keenly in the field of program evaluation. Second, publicly funded programs are to some extent in a state of crisis. After years of geometric growth, programs and those who live by them are suddenly faced with cutbacks in available resources and stringent demands for accountability.

Another impetus for the evolution of alternative perspectives and paradigms comes from the shift in the need for and expectations of evaluation. Evaluation's historical roots are in education. Education is generally concerned with teaching specific knowledge, skills, and competencies that are measurable in some way. A common assumption in education is that specific knowledge, skills, and competencies are cumulative and eventually sum to higher-order outcomes over the course of one's education. Evaluation, in this context, is focused on specific, enabling outcomes as indices of higher order objectives, such as lifetime earnings, personal satisfaction, or contribution to society.

When evaluation was invited into the realm of social programs on a full-time basis in the early 1960s, some different kinds of questions were being asked. Evaluation suddenly was faced with considerably less specificity than it was accustomed to. Mental health interventions were seldom intended to change one specific skill or competency. This was particularly true within the rapidly expanding community mental health movement that flourished in the 1960s and 1970s. In these and related health, education, and social welfare interventions, it became increasingly difficult to talk about what programs do without using some systems notions and concepts. The shift was from a question of outcome to a question of impact.

During the early development of program evaluation, the term *outcome* implied or carried with it the notion that the end result of a program could be specified simply and that this single unitary result could be measured with precision. The shift to the term *impact* to describe the effects of a program, however, brought a far broader and more general meaning to program evaluation. Rather than imply clerical and statistical measurement functions, this concept raised questions such as "impact on whom," "in what environments," and "under what conditions." Implicit in the impact question was the notion that a program is not a unitary phenomenon that is easily and directly measurable. Instead, the term *impact* implies that the program might affect different people, objects, and events in different ways and that program effects may be both indirect and diffuse, affecting some entity in the environment that in turn affects some other entity in a chain-reaction or domino-like fashion. In short, the impact question owns up to the complexity that may be inherent in what a program does. The outcome question implicitly ignores the issue of complexity.

The remainder of this chapter is devoted to elucidating aspects of a systems perspective, which is synonymous with this impact notion of program evaluation. Central to discovering the impact of a program is the idea that programs are best conceptualized as systems. It is toward an elucidation of this view of programs that we now turn our attention.

A Systems View of Programs

"Ex-con Works to Change System." So read a headline in the local papers recently. Of course, it is assumed that the particular system referenced is the criminal justice system, not the educational system, the welfare system, the transportation system, the postal system, the social security system, or the health care system. What about the social system? Well, the article reported this individual's efforts to change the criminal justice system, more specifically, the prison system. But, it is reasonable to expect that a change in the prison system might ultimately affect the social system of which the prison system is a part. It might also have implications for other systems within the social system, such as the political system.

Is it useful to conceptualize our programs as systems? The foregoing example suggests that we already do so, or at least that the general public does. The concept of *system* has become pervasive in our society. The impression of the authors is that this concept of *system* is becoming ingrained in modern man, like a coded engram in the brain. It is doubtful that the concept of *system* was as salient for eighteenth century man. In every day language, system can be defined as a unit of complexity, and certainly the amount of complexity faced by humans from birth to death has increased rather dramatically. Thus, the concept of a system has provided a heuristic of sorts by breaking complexity down into more manageable conceptual chunks.

While the concept of a system is common to us in our role as everyman, it seems to be curiously lacking for most of us in our roles as planners, evaluators, and administrators of programs. It is curious because the concepts of *system* and *program* are so similar. *System* is usually defined as some collection of objects or parts interacting for some purpose. This definition can also stand as a rough definition of what we mean when we say *program*. Certainly a program is usually considered to be some grouping of activities, events, personnel, and equipment that interact in some way to make something happen. That something is usually called program goals.

Similarities between systems and programs are more than semantic and some of our evolving professional terminology suggests that a systems perspective on programs is evolving. Terms like *complexity, networks,* and *interface* are becoming common in discussions about programs. Other terms, like *hierarchy, constraint, variety,* and *regulation,* however, are not.

We have applied the systems concept that has become salient to us in our role as everyman to our work with programs, but we have not developed much in the way of specialized "systems" knowledge, techniques, or methodologies for thinking about programs. In other words, we carry some

common language, commonsense conceptualizations of systems and systematic relationships in our professional work, but we have not developed
much in the way of technical systems expertise to deal with the rich complexity existing in our programs. In particular, traditional evaluation
methodology has left much to be desired in terms of coming to grips with
this complexity. While there have been some interesting initial attempts to
deal with the complexity inherent in our programs, the issue has not been
taken up with much vigor.

Our previous discussion of the various systems ideas and approaches in
the literature illustrates that there is a large body of knowledge about systems
in general. It makes sense to think that this knowledge about systems might
be helpful in thinking about programs, especially given the similar nature of
programs and systems. If programs are conceptualized as a special case of
the general class of systems, it would be an error of omission not to at least
consider the potential of these systems ideas to aid in the conceptualization
and conduct of program evaluation.

We propose that there is such a thing as systems-oriented evaluation,
which stands in distinction to the notions of decision-oriented evaluation
and value-oriented evaluation. Decision-oriented evaluation comprises
those evaluation efforts aimed at providing information to decision makers.
This activity comprises almost all of what evaluation practice is currently.
Value-oriented evaluation emphasizes values, means-end relationships, and
a more goal-free perspective than that of decision-oriented evaluation.
Value-oriented evaluation attempts to determine the standards or criteria
upon which judgments of program merit or worth can be made. With the
notable exceptions of needs assessment and some areas of policy analysis
and planning, there is little value-oriented evaluation evincing itself in current practice. The bulk of this activity consists primarily of philosophical
treatises by Dewey (1939), Scriven (1967), Cooley and Lohnes (1976),
Cronbach et al. (1980), and others. There has been little desire among
evaluators to actually *do* this kind of evaluation. It seems a fuzzy, confusing
issue with little in the way of guidelines, techniques, heuristics, or
frameworks for even conceptualizing the implementation of these ideas.

Systems-oriented evaluation encompasses these value-oriented notions,
but adds a conceptual framework that links value-oriented and decision-
oriented evaluation activities in a meaningful way. Decision-oriented evaluation has traditionally "looked down" into program details for answers to
evaluation questions. The match between the reductionistic philosophy undergirding experimental design and the need to summarize data for decision
makers was a natural. Value-oriented notions address more global issues.
These evaluators "look up" at the values that are "pressing" on the program.

Funding, for example, is a universal constraint on programs. Funding, however, is merely an operational definition of values. Somewhere, somehow, somebody's values were transformed into an amount of funding. Of course, that a program exists at all is a similar expression of value. The match between *all* program effects (not just those incorporated into a program's objectives) and those "higher-order" values is the primary concern of value-oriented evaluation.

Systems-oriented evaluation consists of linking these reductionistic and global perspectives by holding them in mind simultaneously. Whereas decision-oriented evaluation emphasizes parts and value-oriented evaluation emphasizes wholes, systems-oriented evaluation is concerned with part-to-whole relationships. Therefore, systems-oriented evaluation is out of necessity dependent on these diverse traditional and philosophical precedents. Having both perspectives gives the evaluator the freedom to look at *both* program components and contexts. A conceptual framework for guiding such evaluative activity has not been forthcoming, although useful bits and pieces are strewn through the systems concepts and notions we reviewed in the last chapter.

Systems ideas are not new to evaluation, but despite their historical presence and, in some cases, seductive innovative flavor, these ideas have not contributed appreciably to evaluation practice. This result is due in part to the relative safety of using traditional experimental and quasi-experimental designs and providing some individual called a decision maker with various and sundry summaries of the data. This activity largely absolves the evaluator of any responsibility for ascribing meaning to the data vis-à-vis the values that exist in some larger context and from which the program ostensibly derives its rationale and objectives. From the time the report lands on the decision maker's desk, the evaluator can rest easy. The onus is on upper echelon decision makers to convert these data summaries into some meaningful interpretation and action.

Furthermore, in this style of evaluation, the evaluator does not have to do anything very hard. At most he or she may have to coax a statement of objectives from program personnel, operationalize those objectives in some measurable way (usually with an eye toward time and cost), plug these measures into a statistical design that is either well-known to the evaluator or can be learned by one adept at reading statistical texts (after all, for most of us the preponderance of our graduate training was devoted to the acquisition of such skills), and finally generate a report that is readable by some audience. Given the narrow perspective inherent in the decision-oriented stance toward evaluation and the cozy niche this perspective allows evaluators, it is not surprising that the newer systems conceptions of evaluation with their attendent implications for methodology have not stimulated much activity.

We believe a systems view of programs can illuminate some major issues confronting program personnel. For evaluators these issues include political influences, evaluability, bias in evaluation research, credibility and utilization of research findings, quantitative versus qualitative, experimental versus applied, or controlled versus naturalistic arguments, and the like. For program planners and managers, the systems perspective provides a framework for assessing one's context of operation. This perspective emphasizes the delicate balance reflected in the evolution and decay of programs, and provides a rationale for integrating the tasks of program planning, development, and evaluation.

Finally, the systems perspective provides a conceptual framework for those members of the lay public and their elected or appointed representatives who are concerned with the adequacy of federal, state, and local programs in meeting society's needs. At the core of the systems view of programs is the idea that programs exist or take form only because there are values in the world. Programs are merely reflections of values, nothing more, nothing less. Thus, any effort to fully comprehend or comprehensively evaluate programs needs to take into account the processes through which values are identified and transformed into policies that function as guidelines dictating the form, content, goals, and of course, the evaluative criteria of programs.

The perspective we propose is invariant across local, state, and federal levels. Granted, programs vary widely across these levels in terms of budget, scope, and purpose. And, of course, the decision makers, decision settings, and decision consequences can vary widely across the three levels. The view of programs we propose stresses the consistency in the values constraining a program. Values constitute the glue holding a program together, and a systems perspective on programs provides a clear view of all stages of program responsibility (planning, management, and evaluation), regardless of program level or context.

This is not to say that contextual or environmental considerations are not important. Quite the opposite. Programs are inextricably tied to their respective contexts to the extent that the environment (context) defines the program. A virtue of a systems perspective on programs is its equal applicability to program contexts that may be polar opposites in form: federal and local programs, or charity-oriented and profit-oriented program settings, even the "social" programs of tyrannical dictators as well as the programs that evolve under more democratic forms of government.

Programs and the Systems Paradigm

In summarizing the failure of traditional evaluation inquiry to promote an understanding of the programs we create, implement, and live with in Chapter 2, we concluded that short-term temporary solutions to larger-scale problems, poorly thought out funding justifications, and avoidance of personal responsibility by key individuals within systems had evolved as major means for dealing with the myriad complexities inherent in our society's social needs and our values concerning those needs. It was argued that these practices had encouraged the growth of educational systems, mental health care delivery systems, and public welfare programs that are monstrously large, unmanageable, unaccountable, and not evaluable, yet seemingly absolutely necessary for the continued functioning of our society.

Although the field of evaluation has been in the middle of this process, its role is unclear. Most often, it seems to have been responsive to the political realities surrounding its employment. The primary message in Chapters 1 and 2 was that the traditional methodological procedures that had evolved in evaluation were quite consistent with a decision-oriented perspective that fostered the role of evaluation as a pawn in some larger political context. Suchman's (1972) summarization of the misuses of evaluation is illustrative of these denigrated roles for evaluation. Additionally, Brickell (1976), Cronbach *et al.* (1980), Patton (1978), and Weiss (1972a,b, 1977a) provide compelling commentaries on the political abuse of the evaluative enterprise.

Given the failures of traditional evaluation methodology to provide adequate or useful assessments of the behavior of the programs we create, it is only natural to question the status quo. This idea is not new. Guba (1969) made virtually the same statement more than a decade ago.

> When the evidence produced by any scientific concept or technique continually fails to affirm experimental observation and theory arising from that observation, the technique may itself appropriately be called into question. It shall be the burden of my remarks that evaluation as we know it has failed, and that the world of evaluation does indeed require reshaping [p. 30].

Traditional evaluation inquiry has been deeply rooted in the reductionistic philosophy undergirding the use of statistical experimental designs in social research (Riecken and Boruch, 1974; Cook and Campbell, 1975). Parlett and Hamilton (1976) term this the agricultural-botany paradigm, based on the fact that many of these statistical and experimental techniques were originally developed for use in agricultural experiments by Sir Roland Fisher (1925). At the core of this approach is a mechanistic philosophy that attempts through experimental and statistical control to assess the effects of certain variables individually or in small groups while "holding everything else constant."

We suggest that a systems paradigm is a more appropriate philosophical framework for thinking about programs. Rather than attempting to reduce the complexity inherent in programs to somehow make them more understandable or evaluable, acknowledging complexity is essential to fully understanding a program's operation. The reshaping suggested by Guba may be most profitable if the evolving evaluation paradigm that emerges from it is in some ways more consistent with general systems notions and concepts.

A systems view of programs is based on the idea that complexity may be inherent either within a program or as a nontrivial part of that program's context. A systems approach to programs proposes that interactions between variables or groups of variables cannot be done away with, considered linear, or overlooked. Ashby's (1956) admonition that complexity must be accepted as an essential "nonignorable property," is equally applicable to the world of programs as it is to the field of cybernetics. This point of view represents a radical departure from traditional evaluation practice in which the complexity inherent in or around a program is avoided through just such denial and apocryphal assumption. We are in agreement with Britan (1978), Campbell (1978, 1979a,b), Cronbach *et al*. (1980), Glass (1979), Guba and Lincoln (1981), and Koch (1969) that our long-standing empirical efforts to avoid this complexity have not always been worthwhile. Rather than ignore or simplify complexity in order to achieve elegance and parsimony, it may, in some cases, be absolutely necessary to "let complexity in," in order to achieve a parsimonious and elegant statement of program functioning.

Nor are we arguing that our long-standing notions of experimental and statistical control are useless. To the contrary, they are invaluable but only if the results of these methods of control are interpreted within the context of their application, and only so long as the data that derive from them are made meaningful by the explication of the standards or criteria to which they must be responsive. This view is essentially the same as the familiar *gestalt* perspective originally stated by the philosopher Hegel:

1. The whole is more than the sum of the parts.
2. The whole determines the nature of the parts.
3. Parts cannot be understood if considered in isolation from the whole.
4. The parts are dynamically interrelated or interdependent.

This Hegelian concept, in addition to providing some early philosophical impetus for various political and psychological movements, has become a cornerstone of general systems theory. General systems theory strives to find a scientific strategy whereby "internal interactions are left intact and the system is studied as a whole [van Gigch, 1978]." The following list, adapted from van Gigch (1978), relates the potential benefits of a general systems theory approach.

1. The systems approach can be used in considering the relationships of a particular program to its environmental conditions and in identifying factors and variables that affect the program.
2. The systems approach can be used to bring out into the open the inconsistencies of program objectives when treating the various agents that play a part in different programs within the same systems.
3. The systems approach can provide a framework in which the performance of various programs, systems, and the whole system can be evaluated.
4. The systems approach and its attendant methodology can be used to compare alternative programs [p. 30].

In a systems-oriented view of programs, no distinction is made between program and system. Rather, they are viewed as interchangeable terms, both referring to a set of objects, events, or activities that interact for some prespecified purpose. Systems consist of parts that interact to produce more comprehensive outcomes than can be produced by any single part or simple sum of parts, be it in education, mental health, human welfare, medical care, criminal justice, or some other area.

Any particular program is viewed as a subsystem within a larger system which, in turn, is itself a subsystem within some still larger system. Thus, within this view, any particular entity can be seen as a system unto itself, or as a whole that can be broken down into its component parts. This decomposition would represent an *interoceptive* view of the entity. Simultaneously, the entity being examined can be viewed as a component part of some larger entity or process. Viewing this entity within the larger scheme of things represents an *exteroceptive* view of that which is being studied. Through this interrelatedness, change in an isolated part of a system will evidence itself throughout the larger system of which the altered component is a part. Thus, to understand the implications and ramifications of any action, one may need to look beyond the immediate context in which the action occurs.

Any system functions by virtue of the interdependence of its parts. Individual programs are parts of a system and, hence, are meaningful only in relation to other programs in the system and to the goal of the whole system. This view of programs is diametrically opposed to the reductionistic or

incrementalist view that focuses on parts of a system but not the relationships between them. Churchman (1968) comments on the tendency in our
society to subdivide and reduce a system into its functional parts as a
method of improvement.

> It is only natural to expect that improvement can occur in certain sectors in the
> system without having delved deeply into the characteristics of the whole system.
> Thus, for example, there is a tradition in Western thought that parts of the whole
> system can be studied and improved more or less in isolation from the rest of the
> system. So deeply ingrained is this concept of social improvement in Western
> thought that we naturally think it proper that each element develop its own
> criteria of improvement and that the elements be as free as possible from the
> interference of the other parts of the social structure . . . . Man has neglected a
> very serious problem in defining the concept of improvement in large systems
> without understanding the whole system, and if the answer is that we cannot, how
> is it possible to understand the whole system [p. 2]?

Churchman underscores the importance of planning and evaluating in
relation to the context or system in which a program must function and the
pitfalls of any conceptualization that might prevent one from seeing programs as parts or subsystems of larger programs (i.e., from seeing the
means-end continuum).

We would like to provide several examples of what can happen when
planners fail to apply this very simple notion to the programs they create.
The following examples, reported by the Los Angeles News Service (1978),
illustrate the consequences of attempting to improve, in Churchman's
words, "certain sectors of a system without having delved into the characteristics of the whole system."

> Large tractors introduced into Pakistan by Western experts allowed farm owners
> to work more land with less labor, so they bought more land and in the process
> displaced tenant farmers. This put 40 percent of all the farmers in one region out
> of jobs and forced them into urban squalor, with little increase in crop prod-
> uctivity.

> An artesian well drilled on the southern fringes of the Sahara for nomadic
> herdsmen lulled them into staying too long in the area when a drought struck.
> Traditionally, such nomads move at the first signs of drought. This time when the
> herdsmen finally decided to migrate they could not catch up to the receding
> foliage before their herds died.

> The vaunted "green revolution" in which high-yield strains of rice and other
> crops have increased food output has also increased the income gap between big
> and small farmers. The new strains require irrigation and more use of fertilizers
> and pesticides, all of which raise production costs. Large land owners can afford
> it, but small ones, particularly in nations where credit systems are biased against
> them, cannot take advantage of the advanced technology.

> From the Phillipines to Barbados, new mothers have been shunning breast feeding
> in favor of powdered formulas, even though they have no clean water for mixing

them, no fuel to boil bottles and nipples, and no sanitary storage facilities. The babies do not get the nourishment and immunities of their mothers' milk, and the unsterilized and often diluted formula exacerbate the malnutrition and diarrhea that are chronic in Third World infants. Some mothers reportedly stretch a four-day supply to four weeks. Others have substituted corn starch and cocoa for the formula and one Nigerian woman used plain water in the belief that it was the bottle and nipple that provided the nourishment. [Reprinted from "Fitting Technology to Need Held Critical in Third World," by Robert C. Toth, copyright 1978, Los Angeles Times, reprinted by permission, p. C8.]

How are these examples related? The common ingredient among them is that all of the effects described could be attributed to a failure to consider the interaction between a program and its environment. What the "experts" forgot to consider were the interactive effects of their programs on the next larger program or system of which it was a part. The failure to consider this larger context led, in some instances, to an actual decrement in the most significant outcomes of all, even though the programs were successful in meeting their immediate goals. While these immediate goals were being met (i.e., successfully drilling a well in the Sahara, importing tractors to Pakistan, producing high-yield crops, and distributing powdered milk), the larger and most significant outcomes comprising the productivity, self-sufficiency, and health of the people these programs were to serve were not met. How did planners forget to integrate their programs with the long established behavior patterns that had existed for centuries in these regions and that were unmistakable earmarks of the system with which their programs had to be compatible? The answer is, by employing planning and evaluation strategies that reinforced the independence of the component parts of the system and separated them from the purpose of the whole system.

Another example of the consequences of the lack of a systems perspective comes from the work of Hardin. Hardin discusses the effects of the Aswan Dam project in Egypt in which a systems perspective or exteroceptive view of the project would have provided a more realistic estimate of the agricultural, fishing, and medical costs involved. The following is his account of these costs—an account that presents a view much different from others that have praised this project as a "technological marvel."

First, the replacement of periodic flooding by controlled irrigation is depriving the flood plains of the annual fertilization it has depended on for 5000 years. (The fertile silt is now deposited behind the dam which will eventually have to be abandoned.) Consequently, artificial fertilizers will soon have to be imported into the Nile valley. Second, controlled irrigation without periodic flushing salinates the soil, bit by bit. There are methods for correcting this, but they cost money. This problem has not yet been faced in Egypt.

Third, the sardine catch in the eastern Mediterranean has diminished from 18,000 tons a year to 500 tons, a 97% loss, because the sea is now deprived of flood-borne nutrients. No one has reimbursed the fishermen for their losses.

Fourth, schistosomiasis (a fearsomely debilitating disease) has greatly increased among Egyptians. The disease organism depends on snails, which depend on a steady supply of water, which constant irrigation furnishes but annual flooding does not. Of course, medical control of the disease is possible—but that too, costs money.

Is this all? By no means. The first (and perhaps only temporary) effect of the Aswan Dam has been to bring into being a larger population of Egyptians, of whom a greater proportion than before are critically ill. What will be the political effects of this demographic fact? This is a most difficult question—but would anyone doubt that there will be many political consequences, for a long time to come, of trying to do "just one thing," like building a dam on the Nile? The effects of any sizable intervention in an eco-system are like ripples spreading out on a pond from a dropped pebble, they go on and on [cited in Sorensen and Grove, 1978, p. 378].

It is important to note that the significance of these "ripples" cannot be argued away by calling them "side effects." In the common day parlance of planners and evaluators this phrase is often used to refer to the mistakes of planners or the serendipitous effects of programs that seem to have occurred with little or no planning and sometimes in spite of it. As any good planner knows, side effects one day can become main effects another day. Relegating these effects to sideline status is a strategy that assures that they will not be considered as important as "main effects," which the program may or may not be producing. In some and perhaps many cases, the distinction between main and side effects is open to question. Such designations as these serve only to reinforce the mistaken notion that any effect that occurs in some sector other than that narrowly defined by the program's objectives is not as important as a main effect. In fact, labelling some effects as "side effects" is generally a good indication that program objectives were defined too narrowly and the larger context or system was not taken into consideration in planning and evaluating the program. Thus, the concept of "side effect" often serves to bias outcomes in favor of those specified by the program's objectives and to further separate the program from the environment in which it functions. Designating these effects as side effects can never, of course, diminish the planner's responsibility for these outcomes or the evaluator's responsibility for measuring them.

These examples are not unlike many that can be found in education and human service. Nor are they unlike a case that recently came to the attention of the authors in which it was learned after the start of an evaluation that two federally assisted programs were pursuing diametrically opposite goals. One program provided supplementary funds through the county prosecuter's office for police overtime and extra investigatory expenses that might be incurred in securing evidence that would make the case against drug offenders ironclad. The purpose of this program was to reduce the rate at which suspects are arrested but released for insufficient evidence. The additional evidence made possible by these funds often resulted in a stronger case being brought against defendants, thus extending the time offenders served.

A second program implemented was a classroom-oriented program designed to instruct drug offenders in the long-term physical and psychological effects of drug addiction. However, policies governing the use of these funds mandated that only offenders who had light sentences be admitted to the program. New offenders were therefore caught between the two programs in that their sentences might be extended considerably by the first program, making them ineligible for the second program. This resulted in a substantial drop in program enrollees. The solution was to increase enrollment in the classroom program by giving the program to "hard core" offenders. However, these offenders are least likely to be influenced by such a program. Thus, it was no surprise when the evaluation of this program reported its ineffectiveness.

The lack of a systems perspective undergirded by an incremental approach to planning can be noted here too. With the incremental approach programs are planned, developed, and evaluated one at a time outside the context of any larger system within which they must function. Incrementalism is a strategy that reflects the belief that the whole can be improved by adjusting each of its parts independently, as in the case above where the incompatible goals of punishment and rehabilitation were applied within the same system. Incremental strategies have no way of dealing with the possibility that just as one part is "fixed" in one sector of the system another will need fixing in some other sector, sometimes as a direct result of "improving" the first sector. This was painfully obvious in all of the examples above.

It is through processes such as incrementalism that our programs and the activities contained within them may be divorced from higher-order goals, beliefs, and values. This creates the often-found situation in which a program leads to some outcomes that are prized in the short run but inconsistent or noncontributory to other more highly prized outcomes in the long run. In education and human service it is not uncommon to find short-term outcomes confused with long-term outcomes, resulting in the transposition of means and ends in the minds of program personnel. This most commonly occurs when an incremental program strategy is used that focuses on what was referred to in Chapter 3 as "remedial values." Remedial values are usually represented by short-term, immediate outcomes that only partially represent the more comprehensive, longer-term outcomes actually desired in some larger context to which the program is expected to contribute. When remedial values become the focal point for conceptualizing and evaluating a program, program activities and outcomes can easily become separated from and inconsistent with the values that created the need for the program. This is the point at which means can easily become confused with ends.

Some bits of data are in on some programs that indicate that value–program disparity does exist in some of our large federal programs. The economist Milton Friedman reminds us that our present social and educational programs are not immune from the errors of incrementalism and reductionism. Consider the specific examples that follow.[1]

In response to the finding that households of a tremendous number of families existing at or near poverty levels were headed by single mothers, the federal government, between 1967 and 1977, increased Aid to Families with Dependent Children (AFDC) payments to these families by over 500%. Implicit in this move was the idea that these single parents were caught between the need to work and the need to care for their children. Therefore, it was believed that funnelling resources into the family would create the possibility for purchasing day care, meeting costs of transportation to work, providing an enriched home environment, and the like. Furthermore, such a program, in conjunction with other programs like Head Start, would ultimately benefit the children from these families.

Friedman points out, however, that these noble objectives have not brought about intended results. He describes the results of these programs as ranging from disappointing to disasterous. Individuals in these categories have not escaped the poverty trap. This large-scale effort has not significantly altered the economic and social plight of those it was intended to benefit. In fact, our efforts to improve welfare sometimes have served to erode some of the basic values we hold dear!

Friedman argues that these moves on the part of the federal bureaucracy have actually undermined in some recipients the incentive to work and to be self-sufficient. Examples of single female heads-of-households obtaining more real income by drawing AFDC and related benefits than by working are often cited. Here a program with good intentions has undermined a basic value, the right to be self-sufficient. Such programs may foster in some recipients dependency, lack of confidence, self-contempt, and free-floating hostility.

Furthermore, because current rules for AFDC benefits favor single female household heads, they provide financial incentives for marital separation and discourage the formation of family units with both female and male adults present. Thus another of our most basic societal values, that of the

---

[1]These examples were taken from "Free to Choose," a 10-part television series by Milton Friedman presented on Public Television by WQLN Public Communications, Inc., Erie, Pennsylvania. These views are expanded in the book *Free to Choose* by Milton and Eleanor Friedman (1980).

importance of a stable family unit with both parents present, is undermined. Similarly, because the number of dependent children is viewed as reflecting degree of need, these rules provide financial incentives for increasing the number of children in the family. Such incentives undermine the societal values of family planning, population control, respect for human life, and equal opportunity.

So who benefits from these programs? It would seem that often the children in these families, who were to be the primary benefactors of these benevolent efforts, actually take it on the chin. Far from increasing their welfare, these programs reduce their chances for growing up to lead a healthy, happy life. In Friedman's terms, their "freedom to choose" is ultimately diminished by society's efforts to be helpful.

Another example Friedman provides concerns efforts to protect inexperienced and unskilled workers. These groups, with disproportionate representations of teenagers, inner-city dwellers, the handicapped, the uneducated, and especially, the poor and minority groups, are protected by minimum wage laws that ensure that they are paid a fair wage. Friedman criticizes this federal policy, pointing to labor statistics showing that these laws have resulted in higher unemployment rates for individuals whom the program was intended to "protect." Although the law was no doubt beneficial to some people, it is clear that many workers have been priced out of the labor force. Far from "protecting" the young, poor, and unskilled, the law has hampered these individuals' chances of obtaining a job. Unable to obtain employment, they remain untrained and inexperienced, which further reduces their chances of obtaining employment and self-sufficiency. Friedman draws the following picture.

> Here's a person who has low skills. He's only worth two dollars an hour and he would be delighted to work for two dollars an hour. That's the only way he is going to acquire better skills and enable himself to get prepared for a better job. He's a 15, 16, 17, 18, 19-year-old person or a handicapped person, a person who has poor education. No. It is illegal for you to hire him unless you are willing to give him charity, because you can only hire him if you pay him $2.35. It's always been a mystery to me why a young man is better off unemployed at $2.35 an hour than employed at $2 an hour. Can you explain that to me [p. 15]?

Friedman cites other instances of programs clashing with the values and concerns that originally spawned them, including the use of armed guards in desegregated schools as a way of reducing violence between ethnic groups, and the employment of poorly prepared individuals in sometimes thoroughly meaningless and mundane jobs, as has occurred as part of the Comprehensive Education and Training Act (CETA).

Many people may disagree with Friedman's views on economics. However, his opinions on our present social programs, particularly, raise some hard but highly relevant questions for the evaluation of those programs. He holds that many governmental programs "rot away the very fabric holding a society together." His examples of the lack of fit between program objectives and social values, as well as the examples we reviewed earlier, point to one conclusion: It is incumbent on those with program planning, evaluation, and management functions to be very much aware of the part–whole relationships in which programs or program components are enmeshed.

The reader might point out that the examples we have cited are quite relevant to the process of planning but question their relevance to evaluation. They are relevant to evaluation because they exemplify the mismatch between program activities and outcomes on one hand and higher-order values, ideals, and beliefs on the other. Although all the programs we cited could be evaluated as effective or successful, depending on the criteria selected, they often had unanticipated and unintended effects that, in some cases, were quite the opposite of what was intended. This mismatching could be viewed as occurring on a continuum from high to low. It seems safe to guess that programs that produce outcomes highly congruent with these higher-order wants and needs will be considered the most laudable, whereas those exhibiting the greatest discrepancies will be scorned, ridiculed, and often considered a wasteful embarrassment. If there are avenues available to avoid this embarrassment (perhaps through the use of evaluation) in order to save face or save a job, it can be expected that they will be explored.

Do these seem like evaluation issues? They do to us. Far from being irrelevant for evaluation purposes, these examples point to some core issues in the evaluation of programs that have not been addressed adequately by the field. These issues revolve around an appreciation of the program–system relationships that inevitably characterize any program.

It is our contention that the mismatch between values on the one hand and program activities and outcomes on the other is the result of a failure to appreciate the part–whole relationships that characterize a program. The failure to consider the linkage between *global* program characteristics, such as the human concerns that may have motivated the creation of the program, and more particulate aspects of program functioning, such as the measurement of a specific program outcome, appears as a grave omission on the part of the evaluation enterprise.

Evident in the writings of general systems theorists is the notion that the study of systems is necessary in order to understand man's increasingly complex interactions with man and with the environment. Man's increasing awareness of the multidimensionality of the world has led to the search for general approaches by which the human mind can understand events in the context of larger and more meaningful "chunks." The sciences, particularly the behavioral and social sciences, have studied man's relationship to man and man's relationship to the environment in perhaps overly simplistic ways. Recent advances in these fields make it increasingly difficult to understand important concepts and principles without viewing the complex whole of which they are a part. The social and behavioral sciences have, in a manner of speaking, run out of simple problems to solve. Or, more correctly, they have found simple solutions to problems inadequate in the context of recent discoveries and advancements that have all but nullified many of the simple solutions of the past.

Underlying the use of many evaluation designs is the belief that every "confounding" variable must be statistically "controlled" in order to preserve the purity of the evaluation. But this only results in a hypothetical program defined by the absence, in theory, of many constructs that can never in reality be separated from it. This approach is opposed to a systems perspective which studies the manner in which these "confounding" variables relate to or interact with the program in its natural ongoing environment and which accepts these relationships and interactions as essential and nonignorable properties. The folly of the incrementalist view, which attempts to consider parts in isolation of the whole, has been captured humorously in a poem by Beer (1975). Beer ridicules our sometimes tenacious capacity for this type of thinking:

> Take a live thing apart
> To discover what life is;
> You will not find a component called life—
> and behold the live thing is dead.
>
> Shall we take a radio apart
> to find the voice?
>
> —or a car engine to pieces
> to find the speed? [p. 122]

Relationships between a program and the system of which it is a part cannot be done away with, considered linear or negligible, or overlooked, for these relationships are the "voice," the "speed," and the purpose of our programs. Complexity is a fact of life, and complex problem-solving techniques that recognize a multidimensional world are in demand. Simplistic portrayals that deny or ignore multifaceted contextual considerations can quickly lose their appropriateness and credibility when applied to highly complex program settings. Simplistic statistical portrayals, it should be remembered, are honest attempts to provide elegant, parsimonious explanations of the world. However, in the face of spiralling complexity these approaches have traded credibility for their highly valued reductionism.

Needed are some "elegant, parsimonious" ways of conveying the complexity inherent in a program and the system of which it is a part.

It is interesting to note that even within the experimentalist tradition, with its almost exclusive reliance on the study of phenomena in isolation from their native environs, this gestalt view is emerging. In a recent series of articles, Donald T. Campbell (1978, 1979a,b) has posited some notions quite similar to those of both Hegel and the more recent systems theorists. In these articles Campbell acknowledges the problem of interpreting quantified program effects without considering the context in which such quantification occurs. "Qualitative, common-sense knowing of wholes and patterns provides the enveloping context necessary for the interpretation of particulate quantitative data [Campbell, 1978, p. 192]."

In Chapter 2 we noted Boruch and Gomez's (1977) discussion of the corruption and questionable meaningfulness of numbers collected in anything other than laboratory conditions. Whereas Campbell questions *exclusive* reliance on traditional experimental and quasi-experimental designs in social action research and evaluation from a philosophical perspective, Boruch and Gomez's caveats arise from technical statistical considerations. They show in both prose and statistical symbols how environmental complexity might render any particular measure insensitve to treatment effects and thus become a poor indicator of program impact.

Gene V. Glass (1979) is another methodologist who expresses similar concerns.

> Perhaps you will not wish to entertain the following possibility as a "scientific question"; then let me recommend it as a series of "intellectual questions" to be answered by history, political science, and intelligent understanding, as well as by psychology and statistics: Is the vast majority of the variance in education inexplicable in terms of the influences that we currently measure and control? Is it likely to remain so for at least the span of our professional lives? Is that variance, then, essentially irreducible by one who seeks understanding and top-down prescriptive policy about teaching and learning? Is the proper challenge of the disciplined study of education the creation of a system which "satisfices" (to use Herbert Simon's word)? Should our empirical policy studies be based on the assumption that the conditions that make schooling effective are either in practice unknown, unmeasurable, too numerous, or too liable to be controlled by persons at any significant distance from the essential nexus of learning, namely a pupil's brain and a tutor?
>
> I am inclined to believe that the answer to each of these questions is "yes" [p. 14].

Aside from the fact that the writings of these individuals have serious implications for the conduct of program evaluation, the point we want to make here is that the notion of the whole being more than the sum of its parts is not a trite platitude from an esoteric discipline. It is also a caveat being offered by some of our most respected methodologists.

The same general viewpoint appears to be emerging within the specific field of program evaluation. Britan (1978) raises some of the same issues, but his is the most concise juxtaposition of these considerations with the general issue of program evaluation.

> Yet the problems faced by summative evaluation transcend issues of research design and highlight the limitations of an experimental model's underlying assumptions. Summative evaluations treat action programs as if they were laboratories of social change in which specific treatments are always expected to yield particular results. But social action programs are not laboratory experiments; goals are not always distinct, but are often multiple and contradictory; treatments are not single occurrences, but continuing processes; results need not be easily measurable, but are often inexact and multidimensional. Different program treatments not only produce different degrees of success, but different kinds of success, which may be impossible to compare quantitatively. Many action programs aim at qualitative changes in institutions and communities; not easily measurable results, but rather altered contexts for social life.
>
> The experimental model simplifies action program dynamics into an explicit analysis of cause and effect. Often, however, program outcomes cannot be understood without considering the procedures through which action is embedded. Alternative evaluation approaches which consider the complexity of such program settings do not represent watered-down experimental studies, but attempts to understand multidimensional causes and effects through holistic analyses of program context [p. 230; this quote and subsequent quotations cited to Britan (1978) are reprinted with permission from *Evaluation and Program Planning*, 1(3), "Experimental and Contextual Models of Program Evaluation," by G. Britan, 1978, by Pergamon Press, Ltd.].

Virtually all the program evaluation literature we have reviewed and discussed gets at the idea that our methodology simply does not match the complexity inherent in our social problems. Simultaneously we have examined some attempts to deal with increasing complexity in several different disciplines including program evaluation. The earlier program evaluation efforts were almost exclusively short concept papers that tried to elucidate a perspective on how one might go about conducting some alternative mode of program evaluation (cf. Etzioni, 1960; Guba, 1969; Mann, 1972; Schulberg and Baker, 1968; Sjoberg, 1975; Weiss and Rein, 1972), but in very few instances were substantive suggestions provided for how one might actually go about some alternative evaluative course of action. As we stated earlier, despite their bold approach, these ideas stimulated little in the way of a developing rationale for evaluation nor did they spawn much methodology.

Recent efforts, including those of Britan, Campbell, and Glass, just cited do offer more concrete guidelines that could serve to further legitimatize the evaluation enterprise. Generally these suggestions address such issues as consensual validation of results, "triangulation" of data, supplementing quantitative results with qualitative findings, changing one's mental set or assumptions about the world, viewing evaluation as multipartisan education, and the like. These are almost certain moves in the future for evaluation methodology, but even these methodological and conceptual wrinkles do not address the problems of coping with program complexity as directly as necessary. They are more descriptive of current methodological weaknesses than they are prescriptive of solutions to these problems.

Even if we can somehow expand our methodology (and we certainly seem to be doing so), we still need some way of judging when to do what by whom and why. A systems-oriented conceptual framework for thinking about programs would be helpful to evaluators and administrators in making these decisions about program evaluation.

Summary

Inherent in much of what we have reviewed in this chapter is the idea that if some form of complexity constitutes a major or essential program feature, then you should somehow "allow" for that complexity in any analysis or assessment of the program. Since a fundamental characteristic of a program is its organization, the traditional investigation of either its parts or its whole cannot provide a complete description of it. This type of investigation gives us no information about the coordination of parts and process, and of the relationship of parts to the whole. Thus, key tasks of evaluation are the explication of the program in terms of these relationships and their coordination, and the determination of the efficacy of the relationships as specified (i.e., Are they the best means to the desired end?).

The field of evaluation is ill-prepared to contend with relations in a system. Traditional methodology has emphasized quantities of things, with little emphasis given to form, order, and constraint. Thinking about a program's form, order, and interfaces with its environment provides a much different perspective on the concept of *relationship* than do notions of statistical correlation. We do not even have the terminology or syntax to support such a perspective.

So is there no way to keep from drowning in the vast sea of complexity that threatens to swallow any program evaluator who strays off the beaten methodological path? As social science methodologists are wont to do, looking over the shoulder of the hard sciences might provide some direction.

In the hard sciences mechanistic or reductionistic theories once played a particularly prominent role. It was Carl Deutsch who noted that the concept of mechanism:

> implied the notion of a whole which was completely equal to the sum of its parts; which could be run in reverse; and which would behave in exactly identical fashion no matter how often these parts were disassembled and put together again, and irrespective of the sequence in which the disassembling would take place. It implied consequently that the parts were never significantly modified by each other . . . and that each part once placed in its appropriate position with its appropriate momentum would stay exactly there and continue to fulfill its completely and uniquely determined function [cited in Weinberg, 1975, p. 4].

Weaver recounts the breakdown of mechanistic theories in the hard sciences and our change to an interactionist-systems perspective.

> The beginning of the twentieth century witnessed the breakdown of the mechanistic (atomistic) theory even within physics. The laws of physics were insufficient to explain the complex interactions which take place in a living organism and thus had to be postulated—laws of integrated wholes. In view of parallel developments in physics, chemistry, biology, sociology, and economics science became the "science of organized complexity" [cited in Laszlo, 1972. p. 12].

The significance of the breakdown of mechanistic conceptions in the hard sciences should not be lost on the social and behavioral sciences. Even more so than the hard sciences, the social and behavioral sciences need to account for relationships among events. We saw from examples in this chapter that isolating entities into individual parts and, hence, portraying them as mechanisms that are unaffected by laws at any higher level avoids the question of their function in any larger context. The social and behavioral sciences need strategies whereby this larger context is left intact, and phenomena are studied as part of some integrated whole or, in Weaver's terms, as "organized complexity." Only by understanding this complexity can pattern, consistencies, and relationships emerge, allowing for elegance and parsimony in our analysis of man and his environment.

In Chapter 7 we turn to some strategies for studying programs as "organized complexity."

7

A Modeling Approach to Program Evaluation

If we had a way to conceptualize a program such that we could simultaneously see the program's component parts and view the program as a component part of some larger system, we could perhaps gain a conceptual handle on the complexity inherent in that program. The use of what Suchman (1972) called a "rhetorical program model" can serve this function for evaluation. An on-paper representation of a program and its interrelations would be much less difficult to observe, manipulate, and study, than would the *in vivo* program itself. In this chapter we will introduce some syntax for creating graphic models that will help in understanding and communicating the meaning of a program. We will then use these models to pose questions about the program, particularly questions of the part–whole and value variety.

Describing Programs with Models

The heuristic value of a model is well explicated by Pfieffer (1968):

Basic to the entire systems notion is the concept of a model, a simplified but controllable version of a real-world situation which serves a function roughly comparable to that of a laboratory experiment in the physical and biological sciences [p. 6].

The problems confronting us today involve incredibly complex mazes or "nests" of interconnections and linkages rather than straight-forward associations and cause-and-effect sequences, and the results of important changes may be extremely difficult to predict. In such cases a good model can help appreciably by supplementing intuition and judgment. It is one thing to rely solely on speculations about how people will behave under certain real-life circumstances, and something else again when one can objectively investigate possible behavior patterns by simulating those circumstances in a model [p. 6].

A model has an interesting and significant double aspect.... It is an abstraction—a highly simplified version of a fragment of the real world which is too complex for us to deal with directly. At the same time, however, it is one highly effective way of coping with reality [p. 26].

A model is meant to clarify, and to yield information. That depends on how well it is designed. It will certainly be modified or superceded sooner or later in the light of accumulating knowledge, which is the general fate of models. Indeed, from one standpoint the role of a good model is to speed its own absolescence. It cannot provide final answers and is not intended to. It has served its purpose if it provides fresh insights into the working of things [p. 27].

Like all models, including those underlying the use of experimental designs, the program's representation will achieve only a certain degree of fidelity to the real-life situation. Furthermore, there are tradeoffs in the use of models to portray programs. Very simple, easy-to-construct, easy-to-use models, may not be of high enough fidelity to be useful in thinking about programs. Alternatively, models with a high degree of fidelity tend to be very complex and difficult to construct.

Our experiences with various systems analysis and traditional statistical models have suggested to us the use of graphic models for depicting programs. Remembering the old adage that "a picture is worth a thousand words" we have adapted from Ross (1977) a simple *box and arrow syntax* that, we believe, is superior to prose and statistical symbols for communicating some program characteristics, especially (a) program structure and process, (b) the linkages between program activities and outcomes, (c) relationships between values and program outcomes, and (d) relationships between the program and its environment(s). Simply put, some aspects of the way programs are organized make them easy to describe with a graphic methodology.

We will not concern ourselves here with teaching the entirety of this graphic language for describing programs. Our purpose is not to "push" a particular technique. Instead, two simple and basic syntax rules will be introduced to facilitate our present discussion by allowing the substitution of pictures for more cumbersome words. These "pictures" will introduce the notion of communicating conceptually with the aid of graphic heuristics as opposed to relying exclusively on prose.

Here are two simple rules we will use in describing programs:

Rule 1: A box will be used to denote any program transaction, that is, any activity, process, or event. The box means nothing in and of itself but is brought to life or "activated" by inputs (which always enter from the left), constraints (which always press down from the top), and outputs (which always exit the box from the right), as noted in Figure 7.1.

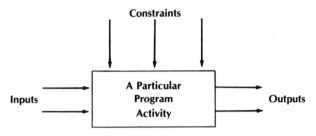

Figure 7.1. **Illustration of a program activity.**

Inputs—things the activity uses; clients or students, tests, staff, time, classrooms, etc.

Constraints—things that constrain the activity; funding, program priorities, feedback on results, anything that moderates or influences output.

Outputs—things "produced" or resulting from the activity; clients with improved outlook, higher rates of employment, students with a certain skill.

Input, constraint, and outcome designations reveal how transactions within a program are tied together. This is illustrated in Figure 7.2.

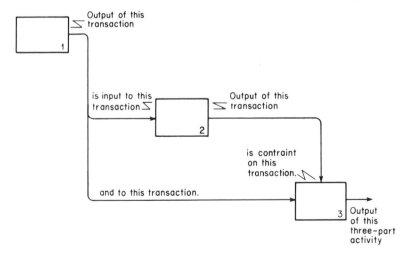

Figure 7.2. **Relationships among transactions.**

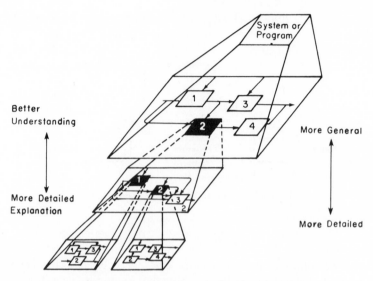

Figure 7.3. **Illustration of a program model.**

Rule 2: Any box, or transaction, can be broken down into its component parts. Alternatively any transaction can be seen as a component part of some larger system. This idea is communicated graphically in Figure 7.3.

The links among program components, or among programs within some larger structure, can be viewed from this graphic perspective in an efficient manner. Considerable prose would be required to communicate all that is communicated graphically, once these two syntax rules are accepted. It is easy with these conventions to imagine (or draw) brief, highly specific transactions within a program component within some program within some overarching national, state, or local policy. In this manner values, the ultimate constraints on what policies will be formulated, are linked to quite specific program details. Of course the model might indicate that aspects of program functioning are not linked to higher-level objectives. Either way, the nature of these linkages is a critical issue for evaluation, but one which has not been addressed because, in part, we have not (as a discipline) had a conceptual framework or methodology for doing so. We are suggesting that these graphic conventions are aids or heuristics for thinking about programs. They give us a way of considering more that is going on at one time than do prose descriptions or statistical models of the same processes.

Models, of course, have some weaknesses too. They make assumptions about the world that may in reality not be true. Models are efforts to simplify and render phenomena understandable, and sometimes they lose fidelity in the process of simplification. Each application of some model will have a corresponding degree of fidelity between the model and the world circumstances it purports to represent. Forcing programs into this graphic systems framework is as liable as any model to these shortcomings.

However, the assumptions behind our program modeling technique differ conceptually from those assumptions underlying the use of our statistical models. For example, the type of program model we propose assumes that (a) *hierarchy* is a highly salient characteristic of programs, and (b) conceptualizing programs in terms of inputs, outcomes, and constraints as the links between program activities is consistent with the way we implicitly describe phenomena. A review of prose descriptions of programs invariably reveals that inputs, outcomes, and constraints are almost universal concepts for describing a program. Such a conceptualization will have higher fidelity in some settings than in others. But, when thinking about programs, we know of no easily readable model that "lets in" more complexity and makes fewer restrictive assumptions than does this type of program model.

Characteristics of the Graphic Modeling Technique

In the pages that follow we discuss the characteristics of this graphic modeling technique that make the models it produces relevant, credible representations of the programs we evaluate. For readers who want to become more proficient in the use of this graphic language the appendix to this volume provides more information and additional illustrations.

HIERARCHY

Man has a penchant for hierarchy. There can be little doubt of that. Quite simply hierarchy is the way we organize things if we wish to consider more than six or seven things at a time. If we get overloaded we create a few higher order categories, each holding a few cases that are similar on some dimension. A hierarchy is a set of objects, activities, or behaviors, arranged according to some pre-established dimension. The characteristics of hierarchical structure are (a) vertical arrangement of subsystems that comprise the overall system, (b) definition of higher-level systems determine the meaning of lower-level systems, and (c) dependence of higher-level subsystems upon actual performance of the lower levels.

A hierarchy, then, is a framework that permits complex systems to be built from simpler ones. On the other hand, hierarchies allow complex systems to be broken down (i.e., decomposed into their component parts and subsystems).

As Simon notes, a complex system is

one made up of a larger number of parts that interact in a nonsimple way. In such systems, the whole is more than the parts not in an ultimate metaphysical sense, but in the important pragmatic sense that given the properties of the parts and the laws of their interaction, it is not a trivial matter to infer the properties of the whole [cited in van Gigch, 1978, p. 375].

Van Gigch (1978) notes that hierarchy implies that:

1. A system is always made up of other systems.
2. Given a certain system, another system can always be found that comprises it, except for the universal system which comprises all others.
3. Given two systems, the one system comprising the other can be called "the high level system" in relation to the system it comprises, which is called the "low level system."
4. A hierarchy of systems exists whereby lower level systems make up higher level systems.
5. The low level systems are in turn made up of other systems and can therefore be considered the higher level system for the lower level systems to be found in it [p. 376].

Simon (1969) explains the nature of hierarchy with the following parable.

There once were two watchmakers, named Hora and Tempus, who manufactured very fine watches. Both of them were highly regarded, and the phones in their workshops rang frequently—new customers were constantly calling them. However, Hora prospered, while Tempus became poorer and poorer and finally lost his shop. What was the reason?

The watches the men made consisted of about 1,000 parts each. Tempus had so constructed his that if he had one partly assembled and had to put it down—to answer the phone, say—it immediately fell to pieces and had to be reassembled from the elements. The better the customers liked his watches, the more they phoned him and the more difficult it became for him to find enough uninterrupted time to finish a watch.

The watches that Hora made were no less complex than those of Tempus. But he had designed them so that he could put together subassemblies of about ten elements each. Ten of these subassemblies, again, could be put together into a larger subassembly; and a system of ten of the latter subassemblies constituted the whole watch. Hence, when Hora had to put down a watch in order to answer the phone, he lost only a small part of his work, and he assembled his watches in only a fraction of the man-hours it took Tempus [pp. 90-91; reprinted by permission of the MIT Press].

Simply put, the concept of hierarchy may be used as a natural ordering or reasoning process to bring order and systematization to the activities, behaviors, or events constituting a program. It appears that hierarchy is our basic means of dealing with complexity. Our communication of complexity invariably involves hierarchy, as evidenced by the structure of written communication expressing more than a handful of ideas. Books are broken down by chapters. Within chapters are major subheadings. Within each of these are paragraphs containing some theme or idea, expressed as sentences consisting of phrases that are made up of specific words. Through hierarchical arrangement a large number of disparate ideas and concepts may be woven into a conceptually understandable and meaningful, albeit complex, piece of communication.

Looking at Figure 7.3, it is apparent that our simple syntax embodies these hierarchy principles that seem prevalent in our society. They are probably prevalent for one reason: They reduce complexity into manageable chunks. No more, no less. If the complexity generated by questions posed to evaluation personnel is overloading the tools and expertise available, it seems timely to suggest some complexity-handling mechanisms, such as the graphic syntax we have introduced.

The hierarchy concept is not necessarily the "true" way the world is, but it might help us understand programs. Two points that further substantiate the role of hierarchy in the world, particularly the world of programs, concern the use of hierarchy as a natural extension of reasoning processes and the inherently hierarchical nature of bureaucracy, the roots from which most programs spring.

Eleanor Gibson (1969) provides a compelling theory of perceptual development that stresses that our increasing ability to extract invariant features in the environment, label the invariance meaningfully, and ultimately make successively finer discriminations is the zenith of our perceptual development. Certainly the ability to use hierarchical structures in one's thinking fosters the act of bringing order or clarity out of ambiguity and chaos, an intellectually challenging task. To our way of thinking, recognizing and using these hierarchical structures makes one a better thinker and problem solver in some instances. The ability to use hierarchical structures to order concepts, events, activities, or data is essential not only to Gibson's theory of perceptual development but to many current theories of human intelligence.

That the concept of hierarchy is somehow engrained in the way we think is perhaps most evident in our organizing behavior. Hierarchy is usually the most salient feature of our organizational charts and appears to be second only to sequence in commonly used flow charts. We will argue shortly that even flow charts depict a hierarchical arrangement of sorts through constraint and that it is constraint rather than sequence that is usually important in describing an ongoing event or process.

Sjoberg (1975) refers to Weber's view of hierarchical bureaucracy and points out its prominence in our society.

> If we look back over the past several decades, we find that the increased bureau-cratization of modern society has been associated with a rising interest in evaluation research. Even today, after substantial criticism on both theoretical and empirical grounds, Weber's conception of bureaucracy, with its emphasis upon rationality, remains a dominant image of how industrial–urban life is and should be organized.
>
> The bureaucratic model rests upon the principle of dominance or hierarchy. Although various positions in the structure are interrelated in complex ways, the overall pattern is one of superordination and subordination. It is within this context of dominance that the notion of universalism is applied in, say, the selection and promotion of personnel. It is within this context that the principle of rationality is applied . . . [p. 34].

Sjoberg's and Gibson's views are introduced as examples of the role attributed to hierarchical arrangements in the way we think and organize in the social sciences. In the "harder" sciences, biological and genetic classification tables and the periodic table of elements from chemistry reflect a parallel prominence of this concept. Even within one's own work environment, such characteristics as *lines of authority, chains of command,* and *superior–subordinate* relationships are usually easily discerned. Yet although this organizing principle is pervasive in our society, hierarchy in-and-of-itself has received little attention.

The modeling technique we will describe is designed to reflect the central role hierarchy plays in the way we organize things, particularly programs. It could be argued that this modeling procedure *assumes* a hierarchical structure that may in fact exist only in the mind of the modeler and not in the phenomena being modeled. As with all models, violations of basic assumptions should render a model useless. However, the pervasiveness of hierarchy in our world, particularly human organizing behavior, suggests that this assumption has a high probability of being isomorphic with the real-world situation being modeled and is a safe assumption for most program modeling purposes. The authors' modeling experiences indicate that if hierarchy is not there, you are not going to find it.

TRANSACTIONS

Some readers might question just what the boxes represent in this modeling procedure. These are program activities or transactions. They are the basic building blocks of programs. The word *transaction* denotes the transfer or communication of behavior or skills from one source to another. Thus, a transaction is any program activity in which the program recipient interacts with program components for the purpose of bringing about some change. Within education, a teacher's lecture, a workbook lesson completed alone by the student, or an arithmetic game played by the class may be viewed as program transactions. Each brings the program recipient (student) into contact with various program elements (lecture, workbook, game) and each has some specifiable outcome.

Logically grouping specific transactions creates more general transactions. For example, a series of teacher lectures or presentations on addition, coupled with the students' working through a series of addition exercises in a workbook, and the involvement of the class in arithmetic games and drills constitutes a general level transaction called "teach addition." This general level transaction is itself but one component part or transaction of some yet more general program activity or transaction, perhaps "teach mathematics" or "teach the 3 R's."

We chose an overly simplistic example to illustrate that transactions are program activities conceptualized at varying levels of generality–specificity. In short, transactions are "two-faced." Any transaction is simultaneously a *whole,* or collection of its subcomponents, *and* a component part of the next more general level of program functioning. Virtually all the programs of which we are aware can be easily conceptualized in terms of their transactions, and further, this conceptualizing can be done at any level of generality–specificity that exists in the organizational hierarchy of a program. Thus, any program component is a transaction that can be viewed interoceptively and exteroceptively.

CONSTRAINTS

We have defined constraints as those arrows that press down on a transaction. We intend to show in the discussion that follows that the concept of constraint is an even more basic and pervasive concept than is hierarchy in the functioning of programs. The shape and form of any given hierarchy is dictated by the constraints and controls that "press" on it. In other words, hierarchy is but a framework indicating how constraints are operating.

Previously we defined constraints as things that affect an activity or transaction, such as funding, program priorities, feedback on results, and other variables that influence or moderate the outcomes of any given transaction. The careful reader probably questioned the definition of a category of events that included "anything that affects output." This is indeed a broad category that demands some explanation. We will first present some of the more mechanistic characteristics of the way constraints operate in the world, and then deal with relationships between constraints, values, and evaluation.

The mechanics of constraint. A constraint does not hold back or limit in the usual sense of the term. Instead, it moderates, influences, or dictates what will happen in one of two ways. *Exteroceptive* constraints indicate "presses" on a program from somewhere higher in the system (hierarchy) than that level at which the program constraint is operating. *Interoceptive* constraints originate at the same level in the hierarchy. For example, funding, which may originate out of policy formulated in Washington, D.C., may have considerable impact on program activities implemented in local settings throughout the country. Consider the example of a federal funding policy in education that provides ample funding for classroom materials, but leaves it to the local districts to supply the personnel to implement the program. Different districts can be expected to be differentially favorable to committing the time of teachers and teacher aides to implement the program. Thus, two exteroceptive constraints that significantly determine what actually happens in the classroom (i.e., constrain the program) would be federal funding restrictions and local staffing patterns.

A real-life example of an exteroceptive constraint comes from the field of bilingual education. Federal law requires school districts to provide bilingual education services to those children with a demonstrated need. Thus in districts with large numbers of this categorical group, the resources available for bilingual education are spread evenly across all those students in need. Some students, however, are particularly handicapped in an exclusively English-speaking environment while others may have equivalent skills in English and Spanish (or French), but still meet the conditions of demonstrated need. The law requires that all be served equally and does not permit local discretion in intensifying the program for some students and attenuating it for others.

Operating simultaneously and independently is the fact that federal desegregation laws require that classrooms reflect the ethnic mix of the community, resulting in classrooms made up of students with tremendous diversity in their language needs. Thus these two *exteroceptive* constraints greatly affect what happens in the bilingual education programs, for they dictate that bilingual education be implemented in classrooms where only some students have a demonstrated need, and even those in need vary considerably in terms of the appropriateness of the bilingual "treatment" being offered.

Suppose the hypothetical educational program example we cited earlier was in actuality the third of a three-part accelerated reading program for gifted students and that the three parts were sequential. Then, as Figure 7.4 indicates, the outcome of this program transaction will be constrained by each student's outcomes from parts I (word recognition) and II (vocabulary). In the manner indicated in Figure 7.4, the outcomes of parts I and II of the

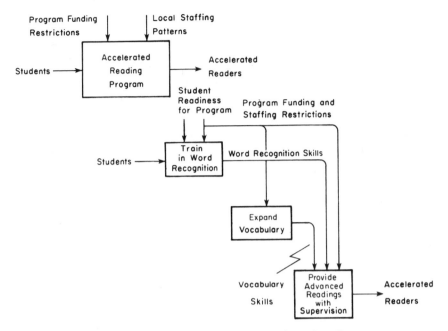

Figure 7.4. **A partial decomposition of an accelerated reading program.**

program interoceptively constrain or moderate the outcomes of part III, supervised reading practice. These are considered interoceptive constraints because they originate from within the program itself at the same level of generality as the program transaction they are constraining.

Another very important characteristic of constraint embedded in Figure 7.4 concerns the *sequence* of program components. Figure 7.4 indicates that word recognition and vocabulary training components must precede practice in reading. The figure says nothing about the sequential relationship between word recognition and vocabulary training components themselves. Presumably these could be implemented in either order.

The related but separate issues of sequence and constraint are often confused in our thinking about programs. Consider the general belief that the ability to subtract and multiply must precede the ability to divide. Out of this belief we sequence training such that students can already subtract and multiply before we attempt to teach division. The sequence is thought to be important. It's not, really. What is important are the *constraints* on learning division, two of the most salient of which are the ability to subtract and multiply.

Constraint and sequence so often go hand-in-hand that we are cognizant only of sequence, when actually it is the principle of constraint that is critical. Sequence is so salient to us time-conscious humans that the principle of constraint becomes obfuscated. Our reductionistic cause-and-effect paradigms, which depend on time relations and describe phenomena in terms of if-then propositions, are undoubtedly an outgrowth of our obsession with sequence and our neglect of constraint.

We are not saying that sequence can be ignored or that cause-and-effect models are useless. We are saying that exclusive reliance on and blind acceptance of these constructs can obscure as well as shed light on the way the world works. If we note that X tends to occur before Y we assume a sequential relationship. What about those less frequent instances in which Y precedes X, or they occur simultaneously? Given the sequential set we have towards these phenomena, would we even notice these deviations? If we did notice, how would we classify these events? What if the interactive effect of A, B, C, and D combine to trigger X first and then Y most of the time, but not all the time? Is thinking about these problems with our usual sequence-oriented framework going to help us discern *these* relationships? Recognizing and describing constraint can be a powerful tool in program evaluation.

A general rule-of-thumb might be that if sequences occur with regularity in programs, we can expect that constraints are at play. If upon careful examination, no constraints are found, order (or sequence) may not be important to the program. The question is how much of the sequencing in the world represents true constraints and how much is artifact, tradition, convenience, and the like. Granted, artifact, tradition, and convenience are constraints in their own right but they are usually conceptually separate from the entities they "constrain." We are interested in those constraints more intimately tied to the program they hold together.

Constraints as values. Thus far, we have stated that interoceptive and exteroceptive constraints determine what a particular program transaction might look like and what outcomes might result from the transaction. The reader is probably wondering just what constraints really are. Enough vague reference to policy, funding, previously attained student skills, and the like; we need to expose constraints for what they really are: *values.*

We stated earlier that constraints represent the "glue" that holds programs together. It is through the mechanism of constraint that any given level of program detail is tied to both the higher-order (more general) transactions occurring "above" it, and to the lower-level (more specific) program transactions "below" it. The concepts of hierarchy and program modeling are ways of identifying and understanding the constraints of a program. If modeling is the pathway to understanding program constraints, then constraints are the avenues toward understanding the values undergirding a program. Exteroceptive constraints, at *any* level of program detail, are but operationalizations of *someone's* values.

We mentioned earlier that funding serves as a universal constraint on programs. The particular level of funding available to a program can serve as an operational definition of the amount of value placed on a program. The more a particular program is valued, the higher the level of funding and the deployment of other resources is apt to be for that program. Less valued programs are assigned a lower priority and therefore will receive fewer resources than their more highly valued counterparts.

Values usually will not appear in close juxtaposition to the program(s) they spawn, for, as was noted in Chapter 3, values are seldom identified, stated clearly, or even considered, particularly in program evaluation. In the best of all worlds, the way we have organized our society allows and guides us to do and make possible those things that we as a people value. But how do we determine what 200 million people value at any one time on any given issue? And even if those values could be determined, how do we determine the implications of these values?

In our society we accomplish these ends through our political system. Through the legislation of laws and execution of laws (and some jurisprudence about the implications of our actions) we move from a value context to an action context. In this manner values are transformed into constraints upon our federal level programs. These constraints are apparent throughout the system as policies, funding levels, regulations, guidelines, and the like. These higher-level constraints in turn interact with local characteristics to produce the nuances of constraint brought to bear on any program activity implemented at any particular site.

We are not saying that all programs emanate from the federal level. What we are saying is that all programs evolve from value to action contexts through some decision-making process (which may or may not be political, depending on how one defines the term). The guidelines, procedures, structure, goals, and purposes that flow from this decision-making process, in turn, serve as the definition of the program. These constraints or program definitions tie together all program activities, inputs, and outcomes.

Constraints define some or all of the following program characteristics: who will be served by the program; how, under what circumstances, and by whom; what is expected to happen; how performance will be evaluated; and so on. In this manner constraints define and draw the boundaries of programs and transactions within programs. If constraints are interpretations of values, then *programs in actuality are operational definitions of values*. They are expressions or reflections of value. Nothing more, nothing less. If one finds a program which aims to facilitate cause X, it is because someone, somewhere, valued cause X. The program might not further cause X as was the original intent; instead, cause Z might be furthered. This might occur as the result of not thinking the program out thoroughly before implementing it. Or, it might occur as a result of the intrusion of someone else's values on the original intents of the program, as in the case of program sabotage by a vindictive program manager or an uncooperative staff, or restrictions placed on program operations or its staff by some governing board or advisory committee.

It can easily be seen that by the time a program is implemented, it may have been altered to the extent that even program *intents* are contradictory to the values originally providing impetus for the program. And we have not even addressed the issue of what the program is actually doing and whether these ends are consistent with program intents or anyone's values. This is, of course, the basis of our objection to solely relying on the decision-oriented approach in which the role and function of program objectives is so prominent.

Similar views have recently been espoused by Morell (1979). Basing his statement on a paper by Brandl (1978), he proposed that

> politicians are not primarily interested in the *outcomes* of social programs. Rather, politicians are interested in the number and diversity of interests and constituencies which advocate a particular position. If enough people advocate a particular course of action, that is good enough to make a decision. The actual value of that course of action, or its effect on the problem for which it was instituted, is of secondary importance. If there is any truth to this analysis, it is small wonder that information on the effectiveness of programs is not quickly integrated into political decision making. By the time such information is available, politicians are being pressed by other constituencies on other matters. What, then, must the evaluator do in order to be relevant and useful? The answer is that attempts must be made to influence not only politicians, but *also* the constituencies which lobby politicians. That is a much slower process than feeding information to a small body of decision-makers. It involves the process of bringing about a change in the thinking of large numbers of people—of convincing those people that particular programs may bring about a desired result.
>
> It is not enough to convince decision makers that a particular type of compensatory education will help disadvantaged children. One must also convince those who lobby decision makers to institute compensatory education programs. Valid data on the incidence of child abuse did not bring about child abuse programs. Those programs were brought about by the actions of citizen's groups who believed that such programs were needed. Social research might have hastened the advent of anti-child-abuse programs, but that hastening would have taken place only if concerned groups knew of and believed in that research. Given the need to influence a diversity of people and groups, it is certainly not likely that the results of any single evaluation will be immediately useful. The most appropriate time for evaluation seems to be *before* a new program is set up, rather than after it has been implemented. This implies a considerable amount of foresight, patience,

and planning on the part of evaluators . . . [pp. 162–163; this quote and subsequent quotations cited to Morell (1979) are reprinted with permission from *Program Evaluation in Social Research*, by J. A. Morell, 1979, Pergamon Press, Ltd.].

Heretofore we have not had any aids for conceptualizing the role of values in evaluation. Because traditional quantitative methodology attempts to be value-free, values have not found a niche in the evaluative enterprise. There is a need for the evaluation community to be sensitive to the variety of values held by those individuals with some direct or indirect interest in the program. Given this perspective, the field as presently conceived may not be capable of playing the role of "multipartisan educator" as suggested by Cronbach et al. (1980) in their call for the reform of program evaluation.

"Teasing out" the values underlying the constraints on programs would seem a legitimate evaluation activity. If one does not attempt to understand the value context of a program, then regardless of what other activities are pursued, it can be argued that one is not *evaluating* the program in a very important sense of the term. The modeling methodology and its attendent ideas provide one possible framework for understanding the relationships between values and what a program does. This perspective allows values to assume their proper place in program functioning. Rather than ignore values, we see them as the "guiding light" of programs. Hierarchically decomposing a program reveals its constraints at successively more specific levels of detail. Having identified program constraints, inferring values is a less risky conceptual leap.

Similar views have been put forth by Cooley and Lohnes (1976). They assert that the value statements inherent in programs can themselves "be analyzed into a set of propositions subjectable to empirical investigation and that failure to perform such analyses in evaluation studies is inexecusable [pp. 9–10]." They argue that the values that have guided educational practice have traditionally been determined by politics and custom and that their validity has not been challenged by educational researchers. They find it curious that value propositions have evaded empirical scrutiny despite educational researchers' heavy emphasis on empiricism. Clear thinking about values in education is considered essential because educational practice is greatly influenced by the value attached to desired educational goals. The alternative to rational inquiry into values is the determination of values on the basis of power, which places the educational enterprise "at the mercy of special interest groups who commend values favorable to themselves as universals [p. 10]."

As noted in Chapter 3 a basic premise of Dewey's notions about values and valuations was that values could be mistakenly viewed as absolutes only if they were considered out of context. When considered *in context,* values lend themselves to elucidation as propositions about real entities and the error of ascribing to them absolute or universal properties is thus avoided. The task of the evaluator becomes one of ascertaining whether value propositions inherent in a program setting reflect only convention or tradition or whether they imply empirically testable relationships between means and ends.

Finally, we want to note that because our perspective is systems-oriented, it is necessarily value-oriented. If one does not consider more generic issues that press on a program, and operates only on the level of specificity at which the program operates, then one is not pursuing a *systems-oriented evaluation* as we are using the term. Using terms like *input, interface,* and *output* does not make something *systems-oriented.* A concern for part-whole relations is systems-oriented, as when we try to link program activities and outcomes to higher-order constraints and values. In the most immediate sense these constraints and values lie in the larger program, system, or organizational unit of which the program is a part.

MEANS-END

A final but related concept embedded within our hierarchical, systems view of programs is the *means-end continuum.* We mean the term precisely as we did in Chapter 3 and as did Dewey (1922, 1939). He made no absolute distinction between means and ends. Any program activity or condition can be viewed as occupying space on a continuum such that it is simultaneously an end to those events and conditions that preceded it and a means to those that follow. For example, a therapist encouraging a client to improve his or her appearance is a means to an immediate end, improved appearance. Improved appearance, in turn, may be a means to more self-confidence, leading to more sociable behavior, which in turn might lead to a more positive self-regard, and so on. Dewey (1922) made the assertion that only when an end is conceptualized as a means is it fully understood, appreciated, or even obtainable.

These are some rather profound ideas. To say that some end is not even attainable unless its subsequent role as a means is realized does not bode well for the reductionistic, "here-and-now" conceptualizations of most of our program outcomes, particularly our measures of those outcomes. Does this mean we need to think things through more thoroughly or that we must appreciate the fact that you never do just one thing with any program intervention? The implications of Dewey's reasoning are the same as the examples of the mismatches between program values, intents, and outcomes reviewed in the previous chapter.

The systems modeling methodology we are suggesting embodies this concept of means–end continuum just as it does the concepts of hierarchy, transaction, and constraint. The "higher up" one goes in a program hierarchy the more general are the program intents. For example, a cognitive behavior modification treatment package for social anxiety may have as its ultimate goal, increased self-esteem. A more narrow goal would be increased socializing behavior by its participants. At a more specific level of detail the program might attempt to build self-confidence as one means of achieving this end. Within this self-confidence portion of the program may be nested transactions on improving grooming and appearance, teaching self-talk strategies for controlling anxiety, and providing practice by having participants role-play social situations.

The means–end continuum is embedded within the hierarchical structure of a systems view of a program. Our traditional methodology conceptualizes our outcomes only as discrete, unitary events. Thus the means–end continuum, so critical to understanding what a program actually does, has not been a focus of evaluation.

The Use of Models in Program Evaluation

What does this system model perspective have to offer? What will it do for us? We introduced the notions of *hierarchy, transaction, constraint,* and *means–end* as systems-oriented concepts that seem to characterize the way our programs are implicitly organized. We think the modeling we have suggested is useful because models constructed in such a way provide a framework or schema for juxtaposing these basic (perhaps universal) program characteristics with both the content *and* context of any given program.

A model is a heuristic that applies these generic characteristics to a program, highlighting the program's adherence to each of the concepts. Traditional methods do not focus on these program characteristics, but rather force on the program a reductionistic, determinist model that will be applicable in some instances but not in others. While our model assumes some other, qualitatively different kinds of conceptual and structural relationships, it does not force them on the program. The form of the program will dictate the form of the model, not vice versa.

A part of our systems perspective is a decomposition methodology with the following characteristics.

1. It is a graphic language that in precision is somewhere between prose and mathematical portrayals of an event.

2. It allows us to "see" and conceptualize an increasingly complex and multidimensional world.
3. It gives us a heuristic we need for understanding complex problems in our professions.

By now it should be clear that a decomposition model of a program is not like an organization chart of the type used by large corporations to show the relative control, power, and/or prestige of president, vice-president, division managers, unit managers and so on down the line. The process of decomposition represents a *conceptual* breakdown of the program while the typical organization chart represents a *descriptive* breakdown. The latter is less useful for planning and evaluation in that it can hide important conceptualizations needed for understanding and operating a program just as the chart shown to stockholders can hide the fact that it is actually the vice-president, not the president, who may be "running" the company.

Also, decomposition models are not "event" charts describing the administrative stages that participants of a program go through. Instead, decomposition is the hierarchical arrangement of program components revealing a family of interacting subsystems. Interactions can take place both among component parts within subsystems and between subsystems. These interactions represent the ways in which subsystems and their component parts are coupled, or combined into larger components to produce an outcome more complex and more encompassing than could be produced by any single component.

With enough information about program functioning, decomposition models can be made sufficiently precise and detailed to show the conceptual and psychological underpinnings of the program. Decomposition methods represent a detached view of the program such that, ultimately, one may be led to the conclusion that there is no program at all. Hence, users of decomposition methodology should be aware that decomposition methodology will dig beneath the surface impressions of a program to find its conceptual structure, and finding none, will so reveal this fact, regardless of any preconceived notions of what the program is supposed to be doing.

Figure 7.5 presents a conceptual schematic of our program modeling approach. At the most general level, the program is represented as a single box. Values providing impetus for development and maintenance of the program appear as constraints at this level. At the next level of program detail, the larger box representing the whole program is broken into its major subunits, depicted here as A_1, A_2, A_3. The number of subunits and their labeling designations are arbitrary. At the next level of detail, A_1, A_2, and A_3 could be further decomposed. A hypothetical decomposition of A_2 is illus-

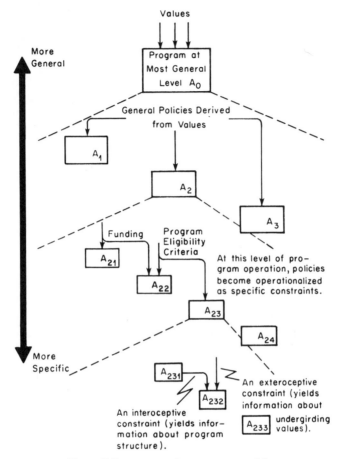

Figure 7.5. **A context-free program model.**

trated. The four subdivisions of A_2 (A_{21}, A_{22}, A_{23} and A_{24}) could be further decomposed and so on.

The values that appear as constraints at the topmost box become successively more specific as one moves down into program details. Values become operationalized with successive levels of specificity in the hierarchy, thus defining important program parameters such as, "Who will be served?" "At what cost?" "In what ways?" "By whom?" "For what reasons?"

At more specific levels of program detail, the transactions represented by boxes become smaller and smaller. In other words, the program "chunks" focused on are smaller and more specific than is the case when one is

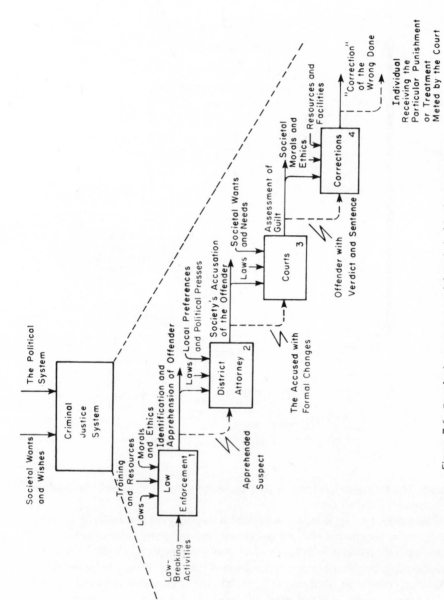

Figure 7.6. **Four basic components of the criminal justice system.**

considering a broader level of program activity. Figure 7.5 illustrates the highly salient role played by hierarchy in this view of programs. Any activity can simultaneously be seen as a program unto itself and as a component part of some larger effort.

The outcomes of transactions at each level of detail should be consistent with and contribute to the attainment of the objectives of the next larger system of which the smaller transactions are a part. It is in this manner that the concept of a means–end continuum is embedded within this modeling perspective. Each outcome or "end" is a "means" to the outcome (or end) at the next higher level of generality in the model. The same is obviously true for constraints. Constraints at a specific level of program detail need to be consistent with higher-order constraints in upper echelons of the hierarchy. To find ends that have no subsequent role as a means to some higher-order end is itself a significant program evaluation finding. The same is true about inconsistencies in constraints at different levels of generality. Such a finding would suggest that very different values are constraining functioning at various levels of program operation, which might hinder overall system performance.

Figure 7.6 presents a high-level view of the criminal justice system. The example is hypothetical but fairly realistic as it is based on personal experience in working with criminal justice systems. In this decomposition the system is subdivided into four primary subcomponents. The primary objective of law enforcement is the identification and apprehension of offenders. This activity is constrained primarily by the laws of the land. These include society's statements about acceptable and unacceptable behavior as well as laws about the treatment of suspects (e.g., use of wiretaps, informing an apprehended suspect of his/her rights). Any law enforcement activity is also constrained or shaped by the training and resources provided to police as well as these individuals' personal morals and ethics (i.e., values) about their work.

Another major activity within the criminal justice system is that process through which formal charges are brought against a suspected offender. In our society this most often occurs through the office of a District Attorney (D.A.) who is usually an elected official. Constrained by the work of law enforcement personnel, public laws, and political presses, a formal accusation is either made or not made. If the case brought against a suspect is weak and without merit or if a suspect's rights are violated in the process of apprehension, charges are likely to be dropped. This is one of the ways the actions of police are an important constraint on the activities of the D.A. In addition, because he is an elected official, the D.A. is pressed to please his constituency. Public concern over a wave of crime or a well-organized effort

by a powerful or vested interest group might significantly affect the functioning of the District Attorney's office.

Society's accusation of the offender (formal charges) then serve as one of
the primary constraints on another major component of criminal justice, the
court system. In addition to formal charges and binding laws, courts are also
constrained by societal wants and needs. Encouraging plea-bargaining to
clear overloaded court dockets is one example of how societal needs may
constrain court activities and avoid a bottlenecked judicial system. Judges
who mete out mild sentences for violations of unpopular laws and the
judgments and punishments meted by juries are examples of how societal
values may function as constraints on the court. Some might argue that
personal preferences and predispositions of the judge (i.e., the particular
judge's values) are an important constraint here as well.

Once adjudicated by the courts, a guilty offender is passed on to the
corrections division of the criminal justice system so that the offender can be
administered the treatment determined by the courts to be fair and equitable
for the offense committed. This "correction" is constrained by societal values about the treatment of offenders and by the resources, training, and
facilities available to correctional personnel. It is sometimes argued that the
values of correctional personnel themselves are a major determinant of what
happens to offenders in the corrections system and some may insist on this
constraint appearing in the figure.

While corrections can take many forms, it most commonly occurs as one
of the four types suggested in Figure 7.7, fines, probation, parole, and incarceration.

Adams (1975) points out that the primary goals of incarceration are (a)
offender rehabilitation, (b) punishment, and (c) incapacitation or detain-
ment. *Someone's* values will ultimately determine whether the *major* thrust
of a correctional agency should be custodial, rehabilitation, or punishment.
Most prison systems contain elements of all three. This conclusion brings to
light some interesting questions. Can we expect a correctional program to
simultaneously incapacitate, rehabilitate, and punish an individual? *Whose*
values are the most powerful constraints on an activity? Are lower-level
outcomes consistent with those higher in the system? These are but a few of
the evaluation issues that are brought to light by this program modeling
approach.

Understanding of a program increases by crossing strata. In moving up
the hierarchy one obtains a deeper understanding of its significance. One
finds broader concepts and longer periods of time, thus allowing the meaning of the program to be more accurately and more easily interpreted. In
moving down the hierarchy one obtains a more detailed explanation of the
program. One finds narrower concepts and shorter periods of time, allowing

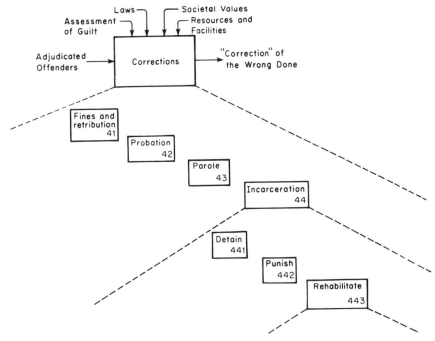

Figure 7.7. **Further decomposition of the criminal justice system.**

for a more precise and detailed explanation of how the system functions to carry out its objectives. Specialization, division of labor, and coordination are three characteristics revealed by a program decomposition.

In summary, there are several common features of decomposition.

1. A higher level unit is concerned with a larger portion or broader aspects of the overall systems behavior.
2. The decision period of a higher level unit is longer than that of lower units.
3. A higher level unit is concerned with slower, more time consuming aspects of the overall systems behavior.
4. Descriptions and problems on higher levels are less structured, with more uncertainties, and are more difficult to formalize quantitatively [Mesarovic, Mako, and Takahara, 1970, pp. 54-56].

Summary

Use of the models we are proposing allows us to see the relationships that may exist in program functioning, not necessarily in a mathematical, quantitative sense but in a conceptual sense. Their use provides a way of seeing more "things" at a time with our limited intellectual capacities. They serve as advance organizers, allowing us to use the information we do possess in a more efficient and productive manner. Use of such models might also allow us to use our traditional statistical methodology more efficiently by teasing out issues and variables conceptually and suggesting ways they may interact to produce the outcomes that do accrue. More appropriate and efficient use of statistics would increase the power of evaluations, not power in the statistical sense (the probability that a statistical procedure will detect a difference that really exists) but power in the sense that the results obtained are important and useful to know.

This modeling perspective represents an attempt to fully understand the context of a program or evaluation problem. We firmly believe that having a conceptual grasp of the context and being aware of the more generic aspects of program functioning makes one a "smarter" evaluator in some ways.

The description of program modeling presented in this chapter represents a rather roundabout way of talking about common sense, intuitive judgment, and evaluation. We have made much of the idea that one should devote some effort to understanding the context in which one toils; by so doing, one is more apt to recognize the role he or she is playing in some larger means-end chain. Knowing this allows one to either better fill the role prescribed or to interject one's own wishes or needs into the work context. Most experienced evaluators delight in pointing out incidents in which they were particularly insightful in applying their expertise. Our experience has been that these instances are often concerned with helping the client or employer to understand needs or appreciate the context in which a program or some personnel operate. Only rarely are these incidents concerned with traditional measurement and experimental design issues.

Although some evaluators already have a systems conceptualization of programs, there are many who do not. Training for evaluators sometimes has consisted exclusively of traditional statistical and measurement methodology. We are advocating the formal inclusion of qualitative or commonsense knowing (Campbell, 1978), clear thinking about issues and events from perspectives other than the cause-and-effect paradigm, and the use of alternative heuristics, such as modeling, as significant aspects of evaluation methodology.

Some would say that we are only talking about a pre-evaluative stage in which one logically thinks through that which will be evaluated. Although they would be right, most evaluators operate from a narrower perspective that does not always provide the evaluator the opportunity of a pre-evaluative stage either because of the myopic view the evaluator has of his or her own role vis-à-vis some program or organization, or because of limitations or constraints placed on the evaluator by the environment. For whatever reason, these constraints limit evaluative insight.

As Adams (1975) points out, a systems or "common sense" perspective can take several forms. Thus, such an approach might be a highly formalized systems analysis methodology that may or may not include a graphics syntax, mathematical models, and the like. Alternatively (and more comfortable to us), this systems perspective might be pursued more informally. Rather than a highly visible process, one's "systems analysis" might be performed by the systems-oriented evaluator within the privacy of his or her own thinking about a program. Regardless of the mode pursued, the critical issue is whether the program and its evaluation are understood, particularly within its larger enveloping contexts. We offer our modeling perspective as one means for achieving such an understanding. For readers who wish to learn more about this modeling approach, we recommend the appendix to this volume.

8

Systems-Oriented Evaluation

Adopting a systems view provides a much different perspective on program evaluation than do more traditional paradigms (experimental design, organizational charts, internal versus external, formative versus summative, and the like). A systems-oriented evaluator is someone who knows the "two-faced" nature of program transactions and is willing and able to look both "up" and "down" from that point in the program hierarchy.

By "looking up" we mean that the evaluator is cognizant of the constraints pressing on a program and is predisposed to questioning what these constraints represent. By looking up the evaluator sees the broader context in which the program unfolds and understands the means–end continuum. The means–end continuum, in turn, suggests to the evaluator standards and criteria upon which an evaluation of the program can be based.

In "looking down" or focusing internally on the program's components, the evaluator is predisposed to performing any of three fairly common program evaluation functions. The evaluator may focus on the success or failure of the whole program to achieve certain prespecified terminal goals (usually called summative evaluation). Alternately, the evaluator may focus on determining the efficacy with which certain program components elicit enabling outcomes and thereby determine the need for modification, deletion, or revision of these components (usually termed formative evaluation). A third evaluative function inherent in looking down into a program is process evaluation, which focuses not on the outcomes of either the whole program or component parts, but on the description and documentation of program activities or processes. Process data often include such information as the amounts of time, effort, and resources devoted to the delivery of services; utilization rates (descriptions of who used the program's services); and documentation that meetings, treatment sessions, and the like have occurred or that certain materials, lessons, or exercises have actually been presented to program participants.

Looking down into the program is a fairly comfortable activity for evaluation; these activities constitute the "stuff" of traditional evaluation. The question of whether our reductionistic methodology fostered this view or whether looking down facilitated our adoption of reductionistic methodologies is of some historical interest. We merely point out, however, that our traditional methodologies and our predisposition to look down go hand-in-hand, and the evolution of this association probably cannot be attributed to serendipity or coincidence.

Why has "looking up" not evolved naturally and practically, as well? Several reasons suggest themselves, but the most central concerns the role of evaluation vis-à-vis management. Looking down is a statistical and clerical function often assigned to evaluators by management. Looking up, alternatively, is usually considered a management function. Consider the following issues. "Are we doing what we intended to do?" "Have we optimized allocation of resources to meet some end?" "How well are various program elements or activities being coordinated?" These are clearly evaluation questions *but* they are questions managers usually ponder, not evaluators. Evaluators may be asked to supply some particulate data bearing on these issues but often this is as close as the evaluator is likely to get to these questions.

This distinction between the evaluator's and the manager's roles may exist because these higher-order evaluation questions are much more related to *personal* accountability than are the more traditional evaluation questions, which usually address *program* accountability. Managers or decision makers are more likely to trust more "personal" answers to these more "personal" questions. They are not as likely to trust an evaluator–statistician for providing answers that may address these personal accountability and management issues. As Attkisson, Brown, and Hargreaves (1978); Broskowski and Driscoll (1978); Weick (1969); and others have pointed out, most theories of management and organization include evaluation as an essential part of the process of decision making. We are suggesting, however, that there are two distinct levels of evaluation that have traditionally occurred in programs. *Program evaluation* consists of looking down into a program to determine its accountability, to find ways of improving program functioning or to document program activities. *Management evaluation* is more concerned with how program results match the intents, wants, and wishes of program actors higher in the program hierarchy.

That these two levels of evaluation have evolved separately is not surprising. First, evaluation is a fledgling discipline that has not "shown its muscle" much to managers and decision makers. Second, evaluation of programs is usually *mandated*. It is something many program managers *have* to do. It is human nature to resent that which has been forced down one's throat. A third reason concerns the person who fills the role of evaluator. If a person with an undergraduate degree and little evaluation experience is hired as an evaluator, he or she will not be expected to understand the management issues at hand, much less be entrusted with providing answers to higher-level evaluative (management) questions. On the other hand, if a Ph.D. level evaluation person is hired, he or she may be *too* understanding of these issues and, hence, not be made "privy" to higher-level management functions and thus higher-order evaluation questions.

Much has been written on the role of the evaluator in the organization (cf. Ashburn, 1973; Attkisson, Brown, and Hargreaves, 1978; Wildavsky, 1972). A continuum of functional roles for the evaluator prepared by Attkisson *et al.* (1978) appears in Figure 8.1. These writers as well as most others (Patton, 1978; Wildavsky, 1972) argue that for evaluation to be utilized and meaningful, evaluation functions need to be integrated into the management function. Attkisson *et al.* (1978) propose methods through which this linkage might be achieved. Similarly, Ashburn (1973) and Wildavsky (1972) propose ways in which evaluation may be integrated or institutionalized within an organization.

Our position, however, is that the *field* of evaluation (its methods, procedures, and expertise) has not had much to offer managers. In some cases, evaluators with considerable savvy and expertise may become managers. In these instances evaluation expertise comes to bear on higher-order evaluation problems. But in the majority of cases, managers and evaluators are distinct groups. The best thinkers in an organization are supposed to be the managers. These individuals are not likely to countenance the assignment of critical management level evaluation functions bearing on their performance to underlings, not even evaluators. A clerk is not likely to be too helpful with these issues. Alternatively, a really competent evaluator might be *too* helpful.

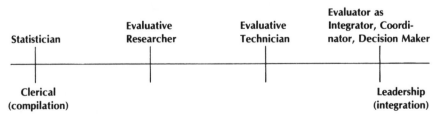

Figure 8.1. **Functional role of the evaluator.**

Similar reasoning has been put forth by Gurel (1975).

A second kind of threat inherent in evaluation may be more immediate but is no less profound in its impact. Simply stated, evaluation poses a serious threat to whatever degree of power the program manager may exercise. Whether through guile or through naivete, the evaluator may seek to pass off his work as a neutral activity intended only to develop knowledge for knowledge's sake. However, program managers, especially if they have clawed their way up the administrative ladder, are not likely to be taken in by the evaluator's assurances, no matter how innocently offered. They recognize that information and knowledge are not neutral quantities once they enter into the public domain. Knowledge is power. Depending on whether the information is withheld and, if disseminated, how and where it is disseminated, information can be made to serve political ends. . . . The point here is that most program managers are probably well aware that they can use evaluation results to suit their own political ends and their own attempts to hold power—and that these results could just as easily be made to serve the opposing political ends and power needs of subordinates and supervisors to say nothing of a host of potential critics and competitors [p. 15].

It should be evident that we are not addressing the role of evaluation in the sense that it is usually discussed. Nor are we addressing the functions an evaluator might perform for an organization. We *are* addressing the perspective the evaluator takes to any role or function he or she is assigned, be it statistical or clerical functions on one end of the continuum or integrative decision making on the other (see Figure 8.1). We believe the perspective one has might in time affect the functions one may serve in an organization but we are not addressing these functional roles *per se*. It is our belief that a systems perspective, which simply means being predisposed to looking both up and down, makes one a better thinker about programs, whether one is an evaluator *or* manager.

Looking Up: The Legitimacy, Representativeness, and Appropriateness of Program Objectives

We have said that "looking down" into program details has been the traditional concern of the evaluative enterprise. It is what evaluators do best and is usually the expertise sought when one hires an evaluator. Evaluation has come to be defined as technology for determining whether a program or transaction within a program is achieving its objectives (Morell, 1979). As indicated earlier, this activity can be pursued in summative, formative, or process modes. This has been the usual entry point for evaluation.

Although the act of looking down in evaluation is supported by tradition and history, the act of looking up has few precedents. Looking down has as its usual entry point the objectives of a program. Looking up begins from the same entry point. However, rather than proceeding from a statement of objectives to an assessment of whether the objectives were met, looking up questions the objectives themselves.

In looking down we move from objectives to measurement pretty quickly. Measurement and evaluation have traditionally gone hand-in-hand and, as we stated earlier, the process of looking down and the construction and administration of specific measures have been mutually supportive. The major argument in Chapter 2 was that these mutually supportive processes fostered the view of evaluation as a statistical–clerical function, thereby undermining its role in the planned change process.

Looking up is the antithesis of this subservient role for evaluation. This is all well and good. However, because almost all evaluation has focused on looking down, there are few guidelines or role models for looking up. The field of program evaluation, with the exception of needs assessments, has not demonstrated an interest in looking up and/or has been denied the freedom to do so. As pointed out by Cronbach *et al.* (1980), "Evaluators are not encouraged to ask the most trenchant questions about entrenched programs [p. 5]." Instead, evaluation has concerned itself with the fairly narrow, reductionistic task of determining whether a program is meeting its objectives. Objectives have generally been taken as givens. The only evaluation problems associated with program goals or objectives usually concern their clarity and/or usefulness as criteria for assessing program success. The issue is invariably one of getting a program's staff to supply a clear, unambiguous statement of objectives. Very seldom is the legitimacy, representativeness, or appropriateness of program objectives questioned.

In Chapters 1 and 2 we critiqued the assumptions underlying the use of statistical designs in program evaluation activities. In addition to the usual assumptions underlying the use of parametric statistics, there are the additional assumptions part-and-parcel to the process of applying these procedures in field settings (cf. Boruch and Gomez, 1977, and Chapter 2, this volume). Even more generic, however, are the three concurrent, parallel assumptions made about the legitimacy, representativeness, and appropriateness of program objectives. While these assumptions overlap with those associated with experimental design, they independently and simultaneously contribute to a lack of clarity, credibility, validity, and utility of evaluation results. These assumptions are now elaborated in light of the systems concepts presented in previous chapters.

LEGITIMACY

In the discussion of response validity of evaluation outcome measures in Chapter 2, the primary issue was whether the response measure tapped precisely the domain of behavior change intended by the program. However, these intents, usually called goals or objectives, are themselves assumed to match the structure and content of the program and to follow logically from some undergirding theory or rationale. Assuming that the response measure is a valid measure of program objectives carries with it the parallel assumption that program objectives can logically be expected to accrue given the nature of the program and the philosophy behind it. In their discussion of response validity, Boruch and Gomez (1977) point out the need for a "rhetorical program model" that depicts assumed interrelationships between program components and objectives. They consider such a model to constitute a rationale or "minitheory" for the link between treatment and response.

However, in some cases, program objectives have become entities unto themselves and are taken as givens. The rationale for how the program is to obtain its objectives may never be specified. For the evaluator, the task is often defined as one of finding a response measure congruent with a particular objective. If the objective could not logically be expected to occur as a result of the program, then the power of any statistical design employed in the evaluation of that objective is further undermined, regardless of the measure selected. The probability of detecting the *true* program effect, if indeed there is one, is smaller than a statistical power analysis would suggest.

Methodologies for laying out logical contingencies between components and outcome measures are not well-formalized, although some do exist. Wholey, Nay, Scanlon, and Schmidt (1975), for example, propose the use of a model depicting assumed linkages between program activities and objectives. This model serves as a basis for establishing the evaluability of a program. Churchman (1979) suggests the use of structural equation models as a means of evaluating implementation of innovative educational programs. The models he proposes, however, could be modified to be useful for purposes of examining "logical contingencies" in programs. The Wholey *et al.* (1975) and Churchman (1979) models are graphic portrayals of the way key elements of a program (such as textbooks, teacher discussions, class demonstrations, therapist contacts, receipt of funds by a client, etc.) are linked to program outcomes. These linkages, usually indicated by arrows in these models, are actually hypotheses about the way the program works. Once "teased out," these hypotheses can be tested more quantitatively.

The graphic modeling methodology suggested in the previous chapter appears to be particularly well-suited to these purposes. A strength of this particular technique is the linkage of *all* program activities, inputs, constraints, *and* outcomes (both enabling and terminal outcomes). This modeling procedure allows for greater specificity but does not demand it. Furthermore, the modeling technique we propose is better suited for purposes of exposition and description. Because the procedure can be implemented thoroughly in order to map as completely as possible all the procedures going on in a program, the model produced can be useful as an investigative or diagnostic device for determining the legitimacy of program objectives.

A graphic layout of how the program intends to accomplish its objectives would be helpful to evaluation in addressing several questions about the legitimacy of program objectives. Are program outcomes realistic? Are program transactions substantive enough and arranged such that it would be logical to expect the intended outcomes to accrue? Do some transactions appear "hollow," out-of-place, inappropriate, or antagonistic to program intents? Are there obvious mismatches between program transactions and outcomes (e.g., is a 30-minute film expected to change life-long racial attitudes of a group of employees)? These issues all pertain to the logic of linkages between program components and expected outcomes.

A second category of questions addressed by this phase of "looking up" asks whether the outcomes that actually do accrue are the result of program transactions. Are there "transactions" in the environment (outside the program) that might contribute to the attainment of program outcomes? Could program objectives be expected to accrue independent of program activities? For example, because children are expected to mature physically and cognitively as a matter of course, attributing these changes to a program would have questionable validity. If such outcomes are expected of the program, are there specific transactions in the program that focus on these outcomes? If so, is it reasonable to expect these transactions to produce changes beyond those expected from processes such as maturation, which presumably occur independent of the program? The lack of control groups in most traditional program evaluation designs intensifies the need to address such questions.

We are suggesting that one of the first activities an evaluator would perform when looking up is a questioning of the legitimacy of program objectives. This activity is merely a logical "thinking through" of the program and its intents. It is almost always possible in program evaluation to select some measure that will either be "kind" to the program or be quite "harsh" and "judgmental" about program effects. Although such measures constitute the bulk of the evaluation data reported, they can be quite equivocal and subject to abuse. We advocate the pre-evaluative thinking through of a program as an important preliminary phase of evaluation, preceding selection of evaluation strategies and specific measures.

This logical thinking through (what Stake called examination of logical contingencies) provides the evaluator with a stronger "feel" for the program and puts him/her in a position to produce a less equivocal evaluation. Of course, a less equivocal evaluation may not be what is desired. If this is so, it will become obvious during this first phase of looking up. If this activity is not permitted at all, or if its conduct engenders anxiety, irritability, or anger in management–decision-making factions, this is a cue that an unequivocal evaluation may not be desired. And, this too is important information about the program.

Given the power relations existing in most management–evaluation situations, in the face of resistance the evaluator most likely (and wisely!) would acquiesce at this point and give management what it wants. As Gurel (1975) points out,

> the issue is one of enormous significance. Mental health practitioners do not just offer mental health services, and teachers do not just teach. They believe in mental health and education, often in a particular set of theories or methods, for example, psychotherapy à la Jung or free expression à la Summerhill. Many of the services only recently confronted with the prospect of evaluation are built on sets of assumptions and personal philosophies deriving from the accumulated biases and preconceptions which pass for conventional wisdom. In certain areas, such as mental health services, this assumptive framework can command emotional commitments as compelling as a set of religious beliefs. Only the most intrepid of program managers and service providers could be expected to be other than hostile to an inquiry which could conceivably render totally meaningless their personal and professional investments in these assumptive frameworks [p. 15].

Although Gurel was addressing resistance to evaluation in a more general sense, his comments suggest that resistance to looking up will be particularly strong. But even if resistance completely halts the act of looking up (it certainly has a well-documented history of interfering with an evaluator's looking down!), not all is lost. The evaluator has gained some important information about his or her role vis-à-vis the evaluation problem being addressed. The evaluator is now more aware of the context in which he or she plies the trade, a piece of information that is of considerable value.

Alternately, management might be strongly motivated to "do the right thing" and would have little use for an equivocal self-serving evaluation. Individuals in management may themselves have doubts about the legitimacy of program functioning or of the evaluation process. Given the history of the evaluative enterprise, these more noble, well-intended managers would not expect much help on this issue from an evaluator or evaluation staff. Why? Because evaluation has not had the perspective, methodology, or expertise for helping managers to look up.

Thus, the first step in looking up, addressing the legitimacy of program objectives, is nothing more than a logical thought process. It certainly does not have to be graphic; it can be an intellectual exercise performed introspectively by a program manager, evaluator, or other program staff. Or it can be accomplished through brainstorming sessions with staff or through use of the Delphi technique, nominal groups technique, or other consensus-gaining procedures. However, graphic depiction allows one to see more about the program's theory or framework of action and constitutes a significant heuristic aid in the process of addressing the legitimacy of program objectives.

REPRESENTATIVENESS

Another common assumption about program objectives is that program impact occurs only in those domains of behavior targeted by the program's objectives. The fallacy of this assumption is evident in the examples of mismatches between program intents and outcomes we cited in Chapter 6. Weick (1969) and Hardin (1963) both make a strong case that "You never do one thing all at once. . . . In a highly interdependent system, any action ramifies and has far-reaching consequences [Weick, 1969, p. 106]."

A major point made in Chapter 3 was that *all* program outcomes need to be consistent with the values undergirding that program. Scriven (1967, 1973, 1974) has quite succinctly noted that accepting program objectives or goals as givens is akin to putting on "blinders." Such statements of program intents predispose the evaluator to look only at those program impacts specified by the objectives, while other outcomes of the program, direct or indirect, go unattended and unnoticed. An important premise of Scriven's goal-free evaluation is that *all* aspects of a program's effects should come under the scrutiny of an evaluator and that nothing should be taken as given from the client or agency soliciting evaluation expertise.

The program evaluation issue inherent in the preceding discussion is whether program objectives are representative of all the things a program may be doing. Conceptually, the issue of representativeness is intimately tied to the systems notions of interdependency and means–ends, and its implications for evaluation practice can be rather frightening. Due to evaluation's traditional propensity to derive singular, particulate measures as indices of program impact, evaluators tend to assume that program impact occurs only in those areas suggested by the objectives and only in those domains of behavior tapped by outcome measures.

Given the myriad aspects of behavior any particular human service or educational program might affect and the relatively small number of outcome variables usually tapped by an evaluation design, it seems safe to conclude that we make judgments about programs with much less than complete data. Furthermore, the data selected may be advantageously biased in favor of some individual, group, or interest. Thus, the assumption of the representativeness of program outcomes represents another potential source of bias and equivocality in evaluation results.

The suggestion inherent in this analysis is that if one were interested in conducting a comprehensive program evaluation he or she would look beyond program objectives to determine if any unintended, unavoidable, or ignored program effects, positive or negative, accrue as the program strives to attain its stated objectives. Unfortunately, at present there is little in the way of techniques, methodologies, or heuristics to help identify unknown program effects. However, an appreciation of the systematic relatedness of the world, particularly some feel for means–end relations and cognitive strategies to overcome one's own myopic viewpoint, will facilitate attempts to determine the representativeness of a program's objectives. Consensus-gaining procedures and other attempts to document the best thinking of groups and experts might also contribute to the resolution of this dilemma. The use of graphics as opposed to prose might contribute similarly. At present we can only say that one needs to "get smart" in contending with the issue of representativeness of a particular program's objectives, and that the field of evaluation is in need of some tools, techniques, or procedures for getting smart about this issue. Traditionally we have relied on measurement and hoped that our measures tapped behaviors and events representative of program impact. Our point is that these hopes have rather insidiously become first, assumptions, and subsequently, givens.

APPROPRIATENESS

A final assumption concerns the match between program objectives and the wants, values, and needs of those for whom the program is intended. If program results are not desired, valued, or needed, what difference does it make that the statistical analysis shows the program to be effective in meeting the stated objectives? The objectives in such an instance are not appropriate. The question of the appropriateness of program objectives is intimately tied to the issue of values. A set of objectives (in some form) can usually be derived for most programs. The question can always be raised as to *whose* wants and wishes the program attempts to satisfy. The question of whose wants and wishes the program *should* satisfy and the issue of the appropriateness of a program's relative utility for various groups are highly value-laden, judgmental issues.

We are not suggesting that an evaluator or program manager jump in, feet first, and hand down such judgments. But it does seem relevant to ask who the program is "paying-off," or benefiting, and who the program is "ripping off," or short-changing. Many payoffs are available in any given program. Clients may receive intended benefits. A politician may profit personally through association with some program. Program staffs are paid-off by the continued funding of their programs. A community, governmental unit, or other group can alleviate its guilt or justify other behaviors through affiliation with or support of programs. However, the wants and needs of the diverse groups that are parties to some program will not always be the same. Indeed such consensus is rare. Sometimes program payoffs are even mutually antagonistic, such as when one group benefits at the expense of another. Thus a program may payoff some, while it rips off others.

Friedman and Friedman (1980) make the case that the federal bureaucracy constitutes a quite large and powerful labor union that wields its power to benefit its members at the expense of those whom federal programs presumably serve. The Friedmans cite the favorable salary schedules, attractive benefits, and the virtually ensured permanence of employment enjoyed by this group as outcomes of the access this group has to lawmakers. The Friedmans point out that although the benefits of federal employment have become increasingly attractive and the size of this bureaucracy has increased tremendously, the services it provides have not increased proportionally in quality or quantity. Intended recipients do not receive services, and stories of waste, abuse, neglect, and incompetence in the service delivery system bound.

Of course, many will take exception to these views, particularly those with a vested interest in the continued health and well-being of the bureaucracy (which includes most evaluators!). But let's take a more historical example in hopes of being less controversial. The lands granted and reservations established by the federal government for American Indians provides such an example of a program that quite literally ripped off those for whom program benefits were supposedly intended. The benefactors of this program were legion. Most immediately, railroad and land barons, and countless individuals seeking new lives and opportunities in the frontier were paid off handsomely by this program. Many of us are direct descendants of this good fortune. The Indians, with few exceptions, did not come out nearly so well.

It is much easier to pass judgment on programs 100 years in retrospect than it is to determine the appropriateness of the objectives of present-day programs vis-à-vis the wants and needs of the various parties to those programs. We are not suggesting that evaluators (or any other single individual or group) make these judgments. We are suggesting that attempting to link program activities and outcomes to the wants and needs of various groups constitutes another important element of looking up in program evaluation. However, although the match between program objectives and values seems to be a critical issue for program evaluation, we are in uncharted territory when we begin to address these issues. Strategies and procedures do not exist for determining the match between program objectives and higher-order values or the match between outcomes and these values. Furthermore, we lack even the methodology to determine whose values are guiding the program.

One area where inroads *have* been made is the developing field of needs assessment. Needs assessment has assumed a predominant role in both educational (Kaufman, 1972, 1976, 1977; Kaufman and English, 1979; Witkin, 1977) and mental health evaluation (Bell, Nguyen, Warheit, and Buhl, 1978; Warheit, Bell, and Schwab, 1974). Needs assessment represents an attempt to link program activities to higher-order values by determining the needs of those for whom a program is intended. These data then serve to guide program decision making and activities in order to ensure that the program is appropriately responsive to the needs of those it was intended to serve.

It should be noted that needs assessment studies are usually highly dependent on citizen surveys, indicating a close link to measurement, or on the use of social indicators and/or service utilization data which assume that reducing a large number of cases to a single summary statistic is a meaningful and useful thing to do. Bell *et al.* (1978) point out that these needs assessment strategies all assume

> that the data employed are reliable and valid. In addition, the service utilization approach requires the assumptions that needs are translated directly into demand for services and that utilization data document accurately current levels of needs for services. In the social indicator approach, it is assumed that needs for services bear a monotonic relationship with indices of the quality of life and the social conditions that prevail in the community . . . the basic assumption underlying the citizen survey approach is that information about social, physical and psychological impairments obtained directly from survey respondents or through interviewer's observations is comparable to the information obtained in service facilities by professional service providers [pp. 294-295].

Thus needs assessment strategies provide an avenue for looking up, but they often do so through the reductionistic, quantitative perspective that has traditionally guided the act of looking down.

The focus of the needs assessment literature has been similarly myopic with regard to its context. By focusing exclusively on what is needed by some group or constituency, these procedures shed little light on the values of different parties concerning the meeting of those needs, including those clients to be served. Often needs assessment results are treated as the answer to the question of what the agency should be doing. Although these data are a valuable input, they should constitute but one piece of the puzzle. Indeed, operationalizing needs assessment data is itself a heavily value-laden enterprise.

The issue of the appropriateness of objectives goes beyond the question of what a particular group or constituency needs. Ultimately we must wonder whether program objectives match societal or community wants and

needs. Relevant questions include whether our program outcomes (supposedly guided by program objectives) are means to subsequent ends valued
by society. Were program objectives developed in response to a careful
study of the needs of some particular population, or do they merely reflect
political reality, or the personal interests of some individual, group, or constituency? Experienced evaluators know well that the assumption that program objectives are appropriate can at times be quite tenuous.

The graphic modeling methodology presented in the previous chapter
provides one avenue for pursuing the issue of the appropriateness of objectives in program contexts. Because this methodology strives to juxtapose
program activities with constraints and outcomes, these entities receive considerably more scrutiny than they do during the unfolding of more traditional
evaluation procedures. Our traditional methodology predisposes us to think
of important variables in terms of dependence versus independence, predic-
tor versus criterion, moderator versus suppressor, reliability of measurement,
change scores, and the like. The important systems roles played by these
variables are obfuscated in traditional analyses. Because constraints (and
ultimately programs themselves) are mere stand-ins or operationalizations of
someone's values, the inferences we make are indirectly and unwittingly
about values undergirding a program. Methods that allow variables to play
their systemic roles in program functioning rather than arbitrarily forcing
them into categories convenient for subsequent analysis are likely to yield
information having more fidelity to the real-world situation being studied.

Another important feature of the program modeling methodology that is
relevant to the appropriateness issue is its emphasis on hierarchy. A program
viewed as a whole at one level of generality is broken down into its constituent parts at the next level of program detail. The objectives of these
subcomponents should be consistent with the expected outcomes of the
higher-level system of which the subcomponents are parts. In this manner
the program model provides a criterion for judging the appropriateness of a
program's or program transaction's objectives: their consonance with objectives of the next higher-level program, system, or organizational unit. Of
course consistency of objectives across levels of generality of program
functioning is but one criterion for judging the appropriateness of a program's objectives. However, inconsistencies will always earmark sources of
difficulty and areas of program functioning that impede rather than further
the accomplishment of higher-order intents of the program.

SYSTEMS-ORIENTED EVALUATION AND PROGRAM MODELING

Examining or attempting to examine where program objectives came
from and questioning their legitimacy, representativeness, and appropriateness represent expanded notions of the evaluative enterprise. We are not

suggesting that evaluators per se perform a series of new activities, but rather that the role and function of program evaluation as a process be expected to include looking up as well as looking down. It is only through looking both ways that we can come to understand the part–whole relations that ultimately undergird program efforts. By logically thinking through a program's constraints and noting inconsistencies between lower-order and higher-order objectives, we can more accurately know whose interests are being served by a program. This seems like a meaningful piece of information, if such an investigation is tolerated. If it is not tolerated, *this* is a meaningful piece of information.

We have given primary emphasis in this and the previous chapter to the development of a rhetorical program model to help program personnel look both ways in program evaluation. This graphic depiction of program activities, inputs, outputs, and constraints provides a basic program document from which to proceed. The information needed to develop a program model is no different from the information usually elicited by evaluators at the start of any evaluation. Gurel (1975) points out these information needs and suggests difficulties likely to be encountered as a result of them.

> To begin with, the evaluator must pin the manager down to specific and detailed answers to a list of questions not unlike those that newspaper reporters are taught to ask: who, what, when, where, how and how much. What are the program objectives, both immediate and ultimate; for what target population; what changes are anticipated as indication that the objectives have been met; when and how will changes be manifested; what kinds of activities are supposed to produce those changes; how much change; how are program activities intended to produce change to be conducted; where and for what length of time; with how much and what kinds of resource inputs?

> It is not just that such questions are embarrassing in the sense that they highlight deficiencies in the program planning process—and there almost have to be such deficiencies, since most human service programs are hastily mounted or have grown topsy-fashion over time. The problem is that they are extremely difficult questions, the first question—the issue of program objectives—being the most difficult of all. Yet the evaluator cannot proceed with the design of a meaningful evaluation unless he has the answers; one cannot operationalize objectives into criterion measures, for example, until one knows those objectives.

> Short of a separate paper, it is difficult to convey adequately the extent of mutual frustration and antagonism that can be generated by trying to elicit from program managers an organized statement of the program's ultimate objectives, the major operating goals, the subgoals, the relation of program activities to achieving subgoals, and how some goals, once achieved, will become activities to achieve higher goals. Program managers seem to get fixated at either listing off detailed activities of the program, or, at the other extreme, they offer some vague superobjective, such as helping people to lead more useful lives.

> Some of my evaluator associates have speculated that the difficulty of obtaining statements of objectives reflects influence of covert goals as primary but unstated

objectives. The notion is that managers wittingly or otherwise have hidden agendas which are not made explicit in planning the evaluation as, for example, the real objectives of methadone treatment having to do with cutting down on street crime rather than with rehabilitating heroin addicts. Such explanations may possess some limited validity, especially to the extent that enhancing the prestige and influence of special interest groups constitutes a hidden agenda. However, my own experience suggests that nothing quite so sinister is operating. In fact, rather than anything remotely Machiavellian, I have been impressed with the kinds of terms that recognize the pervasive influence of program objectives on program operation and evaluation. Admittedly, I have not dealt extensively at the very apex of agency management, but, at several levels of management below that, I have found an enormous preoccupation with getting started and moving off, with very little awareness that going somewhere necessarily implies having a series of intermediate destinations leading to an ultimate destination and some plans about the mechanics of the trip [p. 22].

Gurel (1975) subsequently makes a related point relevant to our present discussion.

For present purposes, I will simply repeat that mounting an evaluation requires the most thoroughgoing kind of planning and, specifically, an intensive scrutiny of the motivation of the parties involved. This by no means implies that there is anything sinister about the participants' reasons for initiating the evaluation and that they should be made to undergo "third-degree" questioning. Viewed negatively, all that is being advanced is the admonition that one avoid through better planning the kind of evaluation that nobody wants: the evaluation that ends up addressing the wrong questions, in the sense of questions in which managers are not really interested; the evaluation that is actually a research project in disguise and does not have a built-in assurance that policy implications will be forthcoming; the evaluation whose financial and other resource costs are out of all proportion to the importance of the issues being addressed; the "quick-and-dirty" look-see which tries to answer questions that can only be answered satisfactorily by a much more intensive effort.

General rules of thumb are risky, but I have found it useful to think of the time requirements for an overall evaluation effort as being roughly divisable into thirds: as much time for planning and tooling up as for data collection and an equal amount of time for analysis and reporting. And it is in the very first planning stages that the kinds of questions listed earlier must be explored—all the way from who wants what kinds of information from the study to how that information will be of value and what differences it will make to somebody to have it. Typically, the pressures to get underway with the evaluation are almost irresistible, and they not infrequently reinforce one another. The evaluator, the manager, and the sponsoring authorities may have different reasons for haste, but they all want to see something get started. Too often, the something that gets started is different from the kind of effort that would be wanted by all the concerned parties if the evaluation were adequately planned [p. 24].

We propose that the haste Gurel refers to is generally haste to look down, not up. Our looking up (creating a rhetorical program model, determining where objectives came from) is similar to what Gurel calls planning the evaluation. Both are ultimately concerned with getting beneath the surface to understand the constraints on the program and on the evaluation itself.

Initiating the evaluation process with the use of our program model serves two important purposes. First, it satisfies the presses for haste pointed out by Gurel. *Simultaneously* this modeling process allows the modeler to look up, to question objectives, and to note the relations between objectives, constraints, and values that may be at play in the program. Looking up in this manner allows one to begin the process of familiarizing oneself with the program while at the same time obtaining continual feedback on the acceptability of looking up. The evaluator may be directly or indirectly asked to stop. Alternatively, evaluation expertise may be welcomed by management if it is believed that information so derived will contribute to the resolution of problems associated with the match between program outcomes and higher-order values. Management has traditionally depended on evaluation to look down. Managers have had very little help *from anyone* in looking up. Whether they would elicit such aid if it were available is an empirical question. At this time, it is also moot.

So what does looking up mean? Several things. It means constructing a model of a program in hopes of understanding the program more thoroughly prior to the design and conduct of an evaluation. Such a model then provides a framework that allows one to move up and down freely on the dimension of program generality–specificity. Constructing such a model entails questioning the legitimacy, representativeness, and appropriateness of program objectives. The process of obtaining and synthesizing this information in the model yields as an important side benefit information on why the evaluation was contracted, what is expected of the evaluation, what limits are posed on the freedom of the evaluator, and most importantly, clues as to the standards or criteria upon which the evaluation should be based.

Looking up means being value-oriented, questioning program functioning as an adequate means to some higher order end. Looking up means being able to describe to decision makers the relations between program activities and higher-order ideals and values, and being able to point out inconsistencies in the objectives of different levels of a program, between program components, or between different programs. In sum, looking up means trying to assess as thoroughly as possible the context in which one finds oneself vis-à-vis some program.

Evaluation and Planned Change

In Figure 7.5 (p. 191) we showed a hierarchical view of programs and the central undergirding role values play in evaluation. That figure provided a simplistic portrayal of the way values come to evidence themselves in programs as constraints on specific program transactions.

As Figure 7.5 indicates, values are first transformed into policies. This is usually, but not necessarily, a highly political process. At the next level of detail, these policies appear as more specific constraints on programs. It should be noted that several different sets of policies, and thus potential sources of constraint, may be pressing on a program. The lowest level of detail (highest level of program specificity illustrated in Figure 7.5) depicts structured and conceptual differences between interoceptive and exteroceptive program constraints. This figure does not include or portray the role of inputs and outcomes in program transactions. Instead, it is provided as a succinct graphic depiction of what has been said thus far about values, systems, and programs, that is, that programs are in actuality operationalizations or "operational definitions" of values.

The careful reader will probably wonder how we can discuss the process of moving from values to programs without addressing the process of planning. It could be argued that the perspective offered in Figure 7.5, indeed in this chapter, is more appropriate to program planning than it is to program evaluation. Certainly, the act of moving from values to action involves planning and program planning is traditionally viewed as a conceptually separate entity from program evaluation. We take exception to this dichotomy. It is our view that planning and evaluation are highly interrelated components of the planned change process. The traditional view of the activities as conceptually and methodologically distinct entities has not always been helpful to the development of either, and has hampered appreciation of the planned change process. As a result, evaluators do not possess even the conceptual framework to think about this process.

The planned change process refers to activities pursued by some system to maintain equilibrium. In systems theory terms, planned change refers to efforts to ensure that the amount of variety in our control mechanisms matches the variety inherent in the systems we are trying to control. If variety in the system exceeds the capacity for variety in the system's regulators and controls, the system will tend toward disequilibrium and perhaps destruction. This principle is illustrated by the practice in sporting events of having equal numbers of players on each side. This assures that an equal amount of control and variety is afforded each team. In the game of football, for example, theoretically there is one person to "cover" or counteract the variety that may be generated by his counterpart on the opposite team. Should one player be removed from one team, the game would quickly fall into (in systems theory terms) "disequilibrium" or (in sports jargon) a "rout"—or no game at all. This matching of every source of variety (program transaction) with a comparable degree of control (program constraint) is what keeps the game orderly and potentially interesting. This same principle, called the Law of Requisite Variety (Ashby, 1956), is also what keeps programs functioning effectively. This is what we meant in the top level of Figure 7.5 when we stated that the concept of values creates the boundary or definition of a program. If there were no values pressing down to maintain equilibrium (to define, order, and constrain program transactions), there would be no program.

Applying this principle more broadly, the planned change process refers to our society's attempts to maintain equilibrium. It is generally held that we value peace and orderliness over chaos. We value equal opportunity and basic human rights, including the right to satisfaction of basic human needs such as food, shelter, and medical care. Programs are attempts to serve and satisfy values under the rationale that doing so promotes overall societal welfare and prevents disequilibrium and decay. Thus programs represent one of our major means of control or regulation over the tremendous variety inherent in the diverse American "melting pot."

This simply goes to say that the programs we create as part of the planned change process are responses to problems or potential problems in the functioning of some system within our society. Most commonly they are efforts to return some system, or part thereof, to a state of equilibrium after it has become disadvantageously influenced by forces whose effects were either previously neglected or whose effects would have been difficult or impossible to predict. It is our values, of course, that define for us what is and is not in a state of equilibrium and, hence, the nature of any program. Obviously, if all systems functioned perfectly (i.e., if they always were consistent with our values), we would have no need for a planned change process. However, systems, especially systems in the social and behavioral sciences, do not function perfectly (at least not for long). In Deutsch's (1951) words, these systems do *not* imply that "each part once placed in its appropriate position with its appropriate momentum would stay exactly there and continue to fulfill its completely and uniquely determined function [p. 234]." In other words the programs we create are continually deteriorating and falling into a state of disequilibrium. From this observation has emerged the concept of *open systems* (Rapoport, 1968b; von Bertalanffy, 1968).

Open systems is a designation used by general systems theorists to separate entities that can be influenced by forces outside themselves, such as educational and human service programs, from the more mechanical, closed systems sometimes characteristic of engineering and technology. Thus, a program to counteract a decrease in the math scores of elementary school students, to teach young adults the dangers of drug abuse, or to disseminate medical and dental care to the elderly all may be considered responses that attempt to return some system to a state of equilibrium or prevent it from falling into disequilibrium. Unfortunately, programs that have as their objective to prevent some system from falling into disequilibrium are not nearly as common as programs that have as their objective to return some system to a state of equilibrium. This reduces the planned change process, as it is often practiced, to crisis intervention.

It should be noted that societies are essentially organizations. As Weick (1969) points out, organizations are dynamic entities in a constant state of flux. Thus one can expect values to change continuously. This implies that the planned change process will always be needed in regulating programs.

The planned change process can be assumed to be operating on two planes at any one time. First, this process operates directly through programs to bring about change. Simultaneously the planned change process works in a societal system or planning plane where environmental events, including program outcomes, are interpreted as means to subsequent higher-order ends valued by society. If present environmental events do not appear conducive to achieving valued higher-order ends, interest develops in the planning plane to *do something* to make what is desired happen. This action often takes the form of a program. Figure 8.2 presents a graphic portrayal of the complementary planes that provide the backdrop for the planned change process.

In Figure 8.2 the program is depicted as taking objectives and converting these into outcomes. Traditional evaluation practices, concerned in one way or another with determining whether a program has been meeting its objectives, clearly operate in the program plane. The activities and entities de-

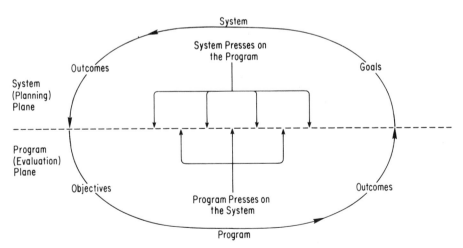

Figure 8.2. **The planned change process.**

picted in the systems or planning plane are more vague, with less in the way of real-world referents to concretize or ground these ideas.

The goals depicted in the system plane are but one example. Somewhere, somehow, the system gets an inkling of what it wants. As Figure 8.2 indicates, existing program outcomes are but one of the presses on the system plane that serve to affect goals. It may be falacious to assume that the effects of program outcomes on goals are as direct as this figure suggests. Other environmental presses not on the "loop" may be more influential.

Weick (1969) points out that organizations and collectives are *assumed* to be rational. He takes exception to the traditional notion that organizing is a means of achieving goal-attainment. Instead of goals determining behavior, his arguments suggest that behaviors provide impetus for goals. He sees meaning as retrospective (i.e., that we infer what we intended—our goals—on the basis of past actions).

> This sequence in which actions precede goals may well be a more accurate portrait of organizational functioning. The common assertion that goal consensus must occur prior to action obscures the fact that consensus is impossible unless there is something tangible around which it can occur. And this "something tangible" may well turn out to be actions already completed. Thus it is entirely possible that goal statements are retrospective rather than prospective [p. 8; this quote and subsequent quotations cited to Weick (1969) are reprinted from Chapter 1, pp. 8, 16, 17 of *The Social Psychology of Organizing*, by K. E. Weick, © 1969, Addison–Wesley Publishing Company, Inc., reprinted with permission].

If Weick's assertion is true, then program outcomes may be most influential in goal generation. This is the same as saying that we determine what we want based on what we have already done.

The "system" depicted in the system plane refers to the planning system, or alternatively, an inquiring system. These terms refer to a process whereby decisions are made that convert goals into specific policies about the environment (the outcomes of the system plane) utilizing and combining the strengths inherent in many different positions and viewpoints. The inquiring system concept reflects a dialectical approach to planning that places trust on informed group opinion. This view of planning considers objectivity in inquiry to result from the strongest possible debate or disagreement on any issue. This process of debate or disagreement is enhanced by bringing together as many different and contradictory viewpoints as possible and considering each to hold an alternative to the problem at hand. Obtaining agreement involves a process of legitimization and consensus by which the ideas of planners, the policies of managers, the needs of clients, and the facts provided by technical experts can be integrated and accommodated to the existing political, social, and administrative system. One design for such a partnership has been posited by Churchman (1971), which he refers to as an *inquiring system*.

The "outcomes" depicted in the system plane of Figure 8.2 are converted into objectives at the program level. The way these outcomes press on or generate a specific program was depicted graphically in Figure 7.5. Thus values are the guiding force in the system plane. The outcomes of this plane constitute general policy objectives (i.e., they are the first or most general level of the operationalization of values). The planning or inquiring system represents the process by which this operationalization is accomplished.

It is in the manner portrayed in Figure 8.2 that programs are born and die, or evolve and decay. Planned change is thus viewed as a continual process. In the same way that Weick (1969) distinguished between organizations and organizing, Figure 8.2 depicts a program as a fleeting moment in an ongoing process to maintain equilibrium in society. According to Weick (1969),

> An organization can be understood only in terms of the processes that are under-way, or its organizing activities, yet it is possible to see regularities in these activities. Organizing and the consequences of organizing are actually inseparable—they are interchangeable notions. The same things are involved and we can call them either organizing or organization depending on how broad a portion of time we observe. Viewing the collectivity for longer periods of time will create the impression that organizing is underway; viewing it for shorter periods will suggest that an organization exists. We observe either an ongoing process that appears "frozen" and steady because it is glimpsed only briefly, or we observe that the process is continuously changing if we watch for a longer span of time. The point is that the crucial events to be explained are processes, their structuring, modification, and dissolving. It is not the tangible fixtures in an organization that are crucial. These merely provide the media through which the processes are expressed [pp. 16-17].

Weick's comments are as relevant to programs as they are to organizations generally. The implication is that programs in and of themselves are not that important. Their role as processes within the context of planned change is important. Thus, planning ultimately represents a larger-order enterprise than does the evaluation of any particular program. We will address the issues of planning subsequently. We conclude this chapter by addresssing another view of evaluation suggested by Figure 8.2 which we call *backward evaluation*.

We have stated several times in this chapter that the evaluative enterprise has traditionally focused on "looking down." This was explained as the act of taking program objectives and through measurement, statistical analyses, and the like, determining whether these objectives have been met. This is all well and good if the program and its relationship to the environment are static, which Weick (1969), among others, asserts is *not* the case. The conceptualization of the planned change process presented in Figure 8.2 suggests that looking down (assessing discrepancies between a program's objectives and outcomes) is a rather myopic perspective for program evaluation to take.

Redefining looking down as *forward evaluation* (moving counterclockwise on the loop in Figure 8.2 from program objectives to program outcomes and determining discrepancies), we suggest that the evaluative enterprise should concern itself more with *backward evaluation* (moving clockwise on this loop from program objectives across into the system plane). A conceptualization of the planned change process provides a backdrop or framework with which one can better come to grips with the implications of both forward *and* backward evaluation. Only within the context of planned change do these perspectives come to light as complementary evaluation procedures. These diverse yet complementary evaluation viewpoints *together* provide a more thorough evaluation of any particular program in the true sense of the word. A simplified view of the planned change process that highlights these dual roles for evaluation is presented in Figures 8.3 and 8.4.

Rather than focus on discrepancies between program objectives and outcomes, backward evaluation has as its goal a statement of the values reflected by the program. Through careful study of a program, the values which that program reflect can be elucidated. Discrepancies between the values reflected by the program and the values in the system plane are of *at least* as much importance as are the discrepancies existing between program objectives and outcomes. Furthermore, because backward evaluation has values as its findings, comparison with value criteria is facilitated. Forward evaluation usually has as its result numerical estimates of some quantity which is not nearly as easy to compare with values; it is analogous to comparing apples and oranges. Backward evaluation circumvents this dilemma.

By doing both forward *and* backward evaluation (i.e., looking both up and down) we gain considerably more information about the state of the

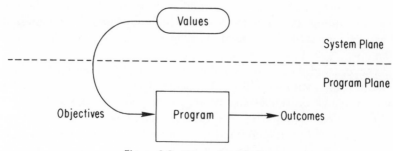

Figure 8.3. **Forward evaluation.**

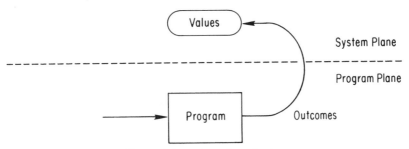

Figure 8.4. **Backwards evaluation.**

planned change process at any one time. Such an approach addresses not only whether a program met its objectives, but additionally sheds light on the legitimacy, representativeness, and appropriateness of program objectives and illuminates the values and interests the program appears to be serving. By doing both, we ultimately realize that in the world of programs we are not dealing with a means-end continuum of program effects. Instead, as Figure 8.2 suggests, the elusive culprits of our scrutiny occupy space on a circle; as soon as we catch it, it has eluded us.

Before leaving this topic we should point out that our views of "looking up" in evaluation are not necessarily novel or unique. The work of Cronbach et al. (1980) raises many of the same issues as well as points out other related perspectives in the literature. Rein's (1976) volume focuses on a critical analysis of stated program goals. Mitroff, Emshoff, and Kilmann (1979) propose an emerging methodology termed "stakeholder analysis." The writing of Gurel (1975), cited earlier in this chapter, is in the same vein. Although these ideas have been around for a while, they have had little impact on evaluation practice.

THE SYSTEMS-ORIENTED EVALUATOR AS AN EDUCATOR

In their call for a reform of the field of program evaluation, Cronbach et al. (1980) suggest that evaluators play the role of multipartisan educators.

Decision making is proceeding well if participants reach decisions with their eyes open. Something is wrong when they support choices they would reject if they saw the likely consequences clearly. Not knowing what each program will accomplish, a citizen may support an attractive option that has little genuine promise. This is wasteful. Worse, it permits the illusion that, because "something has been done," the problem has been vanquished. Those who settle for a glittering promise lose their momentum and their influence. In W. W. Jacobs' story "The Monkey's Paw," a humble man and woman were granted three wishes— with a curse that after each wish was fulfilled they would be sorry they had made it. Regret, less dramatic but ever present, is the lot of participants in a democratic process who throw their weight behind proposals whose workings they misconceive. It is too much to ask for twenty-twenty foresight, but it is the role of evaluation to supply spectacles for every nose, to bring as much into focus as possible [p. 155].

It is our view that being systems-oriented will facilitate the multipartisan educator role Cronbach *et al.* has called for. They describe such an evaluator as one

who is attempting to help members of the policy-shaping community deepen their understanding of a program, a social problem, or the decision-making machinery itself. The evaluator settles for too little if he simply gives the best answers he can to simple and one-sided questions from his clients. He is neglecting ways in which he could lead the clients to an ultimately more productive understanding [p. 161].

These writers go on to state:

If the evaluator does not try to see the program through the eyes of diverse partisans, the posing of the questions may itself—myopically or intentionally— favor the interests of one agency, one sector of society, or the ideology of the evaluator himself. Some lists of outcomes worth measuring, for example, foredoom the program by asking the impossible of it; other definitions turn the evaluator's eyes away from shortcomings and lead to a positive report on practices that are in fact inadequate. Concrete examples of this point have been scattered throughout our book [p. 208].

It is our belief that the systems perspective on evaluation we have offered has some relationship to that suggested by Cronbach and his colleagues. It is only by looking both ways from any particular program vantage point that one may fully appreciate the diverse interests and concerns that abound within the policy shaping community (loosely defined as all parties having some direct or indirect interest in a program).

Summary

This chapter has put forth the notion that programs are operationalizations of values. We introduced in the previous chapter a modeling technique as one way of uncovering the values that undergird a program. We argued for the need to look both up and down in program evaluation, and that looking up heretofore has not been possible or acceptable (in many circles it still is not). We pointed out that an appreciation of hierarchy, means–ends, and program modeling allows one to understand constraint, which in turn is a way to understand the role played by values in program evaluation. In this chapter we noted that both looking up and looking down require the evaluator to examine three often held assumptions in program evaluation. These assumptions pertain to the legitimacy of program objectives (can they logically be expected to bring about the outcomes desired?), the representativeness of program objectives (does program impact occur only in those domains targeted by the objectives?), and the appropriateness of program objectives (do the objectives match the wants, values, and needs of those for whom the program is intended?).

Finally we addressed the role of evaluation in the planned change process. We concluded that working backward (from observing program activities and inferring values) makes as much sense as moving forward (moving from objectives to outcomes), never questioning where the objectives came from. In Chapters 1 and 2 much of our discussion centered around the ills of evaluation. We feel strongly that the discussion in this chapter addresses a root cause of these often-cited ills.

Once the reader understands this, so what? The behavior of a so-enlightened evaluator is not apt to change very quickly as a result of the influx of these ideas. The credibility of an evaluator who asks if he can "look up" or "go backward" is likely to be seriously damaged, even with the most enlightened of supervisors. You may not be allowed, or even want, to do the kinds of evaluation we suggest.

But if you are stopped in the process, either by yourself or others, that is information worth pondering. Even if you never get started with this perspective *in vivo*, you can utilize it introspectively. We have said repeatedly that this perspective helps one understand the context one finds oneself in. In a sense we think our views can make one "smarter" about programs. Evaluators may be limited in the activities they perform, but they are never limited in the ways they may think about programs. Awaiting reform of program evaluation, systems-oriented evaluation will exist as a perspective more than as specific behaviors in which evaluators engage. Hopefully the notions of systems-oriented and value-oriented evaluation will help promote this reform.

9
Conclusions and New Directions

We have covered a lot of ground in the preceding chapters. Although we have presented our major points, there are several ideas remaining. We will use this chapter to address these ideas and to summarize our perspective on evaluation.

Decisions, Values, and Systems: An Evaluation Trinity

While the preceding chapters have explicated the decision-oriented, value-oriented, and systems-oriented conceptualizations of program evaluation, there are some additional aspects of this three-part conceptualization that need to be addressed.

First, we want to acknowledge that these three "perspectives" on evaluation are only theoretical conceptualizations; they do not exist as real strategies or methodologies. Rather than comprising specific evaluation activities and techniques, they are ways of thinking about the "pieces" that constitute the evaluation enterprise. We believe that this conceptualization is a reflection of the primary entities in our tripartite world of programs: decisions, values, and systems. To us it is inconceivable that a program can exist apart from these three considerations. If evaluation is emeshed in decisions, values, and systems, it only makes sense that evaluation should be sensitive to or oriented towards these considerations.

It should also be noted that these three kinds of evaluation do not exist independently of each other. Evaluation activity can always be described as one or more of these three. Furthermore, any particular evaluation context

225

can always be characterized as containing elements of all three. The key to determining which is predominating is in the questions being asked.

Evaluative questions usually reflect a mixture of decision, value, and systems considerations. Asking "Which of several treatments is best?" is a decision-oriented question. There may or may not be a value question inherent in what is meant by best. "What should the agency be doing?" is a question that is primarily value-oriented. The answer to such a question will invariably be highly value-laden. Even this value-oriented question, however, implies some decision-oriented consequences (e.g., what are we doing relative to what we should be doing?).

"How" questions are most often systems-oriented. "How is the program achieving results?" is a systems-oriented question about program functioning. "How are program outcomes related to program objectives, broader policies, and societal values?" is a systems-oriented question about a program's relationship to higher-order systems and goals. Even these questions are not *exclusively* systems-oriented. They are decision-oriented and value-oriented as well.

The point we wish to make is that the kinds of questions being asked will indicate the orientation of the evaluation. Thus, the effective evaluator does not come to the evaluation context with this or that orientation. Rather, he or she responds in a manner consistent with the question(s) being asked. If these questions are primarily value-oriented, for example, then a value-oriented evaluation strategy would be most appropriate. This is similar to Morell's (1979) conceptualization of evaluation as a technology responsive to a wide variety of information needs:

> Evaluation is a form of social research which emerged from a need to test the value of society's efforts at solving social problems. As such, evaluation must generate information which will be useful to decision makers. If evaluation is to fulfill that role, it must operate on a model which is attuned to maximizing solutions within a context of ever-changing political, social and economic constraints. The scientific model with its main goals of advancing truth and theory development, is attuned to surmounting those constraints by developing artificial situations which interact as little as possible with the practical world. Although scientists as individuals (or in groups) may be interested in helping with practical issues, the logical structure of their work is oriented toward goals which are, at best, irrelevant to the development of innovations which will survive the rigors of wide-scale implementation. Hence the need for conceptualizing evaluation as technology, a system in which the reward structure and the use of resources are intimately attuned to the issues involved in developing and testing practical innovations [pp. 118–119].

Another point we would like to make is that virtually all questions (and thus evaluation contexts) will be first and foremost decision-oriented. If there were no decisions to be made there would be no evaluation questions. Granted that evaluation occurs within a decision-oriented context, the question is whether value and systems considerations come into play. Sometimes they will and sometimes they will not. Sometimes the evaluator may be able to introduce them into the decision context as highly relevant issues, sometimes not. To do so might in some cases make the evaluator look like a saint. In other instances it might cost the evaluator his or her job.

Another point we want to make about a systems orientation to program evaluation is that this perspective gives an evaluator more options. In Chapter 2 we put forth the argument that an exclusively decision-oriented perspective is rather myopic in that it limits alternatives available to the evaluator. A systems-oriented conceptualization of evaluation differs radically from the beaten path; this view of the evaluation enterprise draws the evaluator into a broad spectrum of events, activities, and issues that the methodology of experimental design has not permitted. Thus a systems perspective would force some "pruning" of traditional methodology, fostering new growth in the form of new methods more consistent with widely varying information needs.

Systems-Oriented Evaluation as Theory

We have reviewed contentions by a variety of writers that traditional evaluation methodology based exclusively on the experimental method is a bankrupt enterprise. Calls for new techniques and methods abound. It is our contention that the conceptualization or framework of evaluation we have suggested will foster this development in positive ways.

First, our framework legitimizes values and systemic relationships as appropriate subject matter in program evaluation. If these issues do indeed constitute "new ground," then it can be expected that new methodological developments will evolve. Second, the suggested approach provides a theoretical perspective on evaluation that not only "contains" (or has a place for) our traditional methods but provides a theoretical rationale for developing approaches and methods that have thus far achieved only limited utility. The decision-theoretic approach (Edwards, Guttentag, and Snapper, 1975), for example, is a well-explicated methodology for dealing with very thorny problems involving subjectivity, values, consensus, and decisions. However, despite the relevance of this method for practical problems facing many evaluators and decision makers, it has seen only limited use. A systems-oriented conceptualization would facilitate the use of this particular approach as well as other procedures that have a place for values and systemic relationships in their methodology. Thus, the conceptualization we suggest would provide some theoretical bases for already-existing as well as to-be-developed evaluation methods and techniques. Evaluation has continued to grow helter-skelter as a technology in the absence of guiding theory. We believe a systems theory view of programs would provide some direction and explanation for that growth.

Evaluation's growth is not random or serendipitous. Our guess is that, just like always, as new information needs evince themselves, some bright souls will develop ways of addressing those needs. Our point is that these new information needs will have something to do with values, systems, and decisions. And, thus, the methodological response will always be systems, value, and/or decision oriented. Far from random, what we will be doing in the future as evaluators is to a degree predictable. By paving the way for such developments we hope that our perspective will hasten the development and elucidation of new ideas.

The Systems-Oriented Evaluator

Throughout this book we have emphasized that one can be as systems-oriented as one wants in the privacy of his or her own thinking. We also tried to caution that some discretion in being systems-oriented "in public" may be warranted. We would like to sound that caveat again. This book has been an attempt to describe the evolutionary development of evaluation and related fields. Our point has been to show how divergent sources yield similar implications about what evaluation should do to become more a multipartisan, truth-seeking enterprise. We have given our best efforts to elucidating a perspective that would move evaluation in such a direction. It is ironic that the more successful we are in achieving this goal, the greater the resistance will be to the behaviors we are suggesting in program evaluation contexts.

We argued earlier that smart, seasoned evaluators were already systems-oriented; we believe this type of thinking is implicit in the way these individuals approach problems. Along with their poise and social skills, this implicit perspective has helped them win respect as competent professionals. But what about naive, green evaluators? Or, individuals from other professions pressed into service as program evaluators without sufficient training? And, what about students in undergraduate and graduate courses in program evaluation?

While believing that a systems perspective makes one smarter about programs, these individuals, indeed all evaluators, should be cautioned that part-and-parcel with getting smarter about programs come certain risks and liabilities. Blindly adopting one view and letting that perspective determine one's interaction with the program without any thought to the way that behavior is likely to be received by program personnel could be very unwise. In some contexts a systems or value-oriented perspective might be the only way to do a good job and would be welcomed. In other contexts to *behave* as a systems-oriented evaluator might be professional suicide. In other words, taking us too literally might cost you your job. You can certainly *think* in systems terms, but there are places where "looking up" simply will not be tolerated, much less appreciated.

Much has been written about the anxiety engendered in program personnel by evaluation activities. Most of this literature has considered only traditional decision-oriented practices. If these relatively mild evaluation activities engender anxiety, systems practices can be expected to evoke even stronger reactions. What we are saying is that systems-oriented evaluation may be viewed as a revolutionary act; behavior that would label one a traitor in the eyes of some. While we believe that being systems oriented necessarily encompasses the most honorable of intentions, it may not be obvious to threatened parties of systems-oriented efforts. Thus, being systems oriented requires a degree of tact, diplomacy, and professionalism, usually found in competent, veteran evaluators. Being systems oriented in one's thinking is one thing. Being systems oriented in one's behavior is another matter. The world of programs, in some instances, may not be ready for it. The field of evaluation may not be ready either.

Systems-Oriented Evaluation and Doing the Right Thing

Some effort has been devoted of late to the establishment of ethical standards for evaluators (e.g., Joint Committee on Standards for Educational Evaluation, 1981). The topic is being addressed increasingly in the literature as well. Not unlike other treatises on evaluation, this issue is also related to our perspective on evaluation.

First, there is the issue of doing the right thing. This, of course, will mean different things to different people. For us, it means giving as unequivocal and honest an account of the phenomena being studied as possible, unshackled by a program's objectives, viewpoint of program administrators, political consequences, or any of the "presses" that may constrict the activity of the evaluator. Given our view of traditional evaluation as a pawn in some larger political context, it should not be surprising that we view a systems approach to evaluation (as we have defined it) as a way to overcome misuse of evaluation expertise. We believe all the attention devoted to the politicization, utilization, and relevance of evaluation is the direct result of the denigrated decision-oriented role the field of evaluation has chosen to play in our programmatic efforts.

What "good" do evaluators do in this decidedly decision-oriented role? It is our contention that evaluators have contributed very little with this traditional perspective, and we offer the volumes and reams of evaluation information produced in the last 20 years as support for our contention. We began this book by implying that most decision makers ignore or consider lightly evaluation results, especially when it is expedient to do so. We believe little impact and "societal good" have accrued as a result of these efforts. We view the evaluators who produced all of these reports and findings as souls who were trying to do the right thing. Our criticism is not personal. Rather, we are saying that the traditional role we find ourselves playing prevents us from doing a very good job of contributing to the public good, even in an indirect way, and despite the most honorable of intentions.

A systems-oriented perspective, with its emphasis on programs as operational definitions of values, attempts to link what happens in programs with what is wanted, needed, or desired in some larger context. The value-oriented nature of our perspective implies that programs need to be consistent with and build toward some higher-order good. Thus, strong ethical concerns are an inherent aspect of this perspective. Far from ignoring such issues, values, ethics, and "doing the right thing" are an inherent part of the perspective itself.

A second point we want to make concerns the ethical and/or moral dilemmas likely to be encountered by the systems-oriented evaluator. We have already noted that the systems-oriented evaluator is not "playing it safe" when following the role we have described. We suggested that the use of discretion and tact were essential if one wanted to pursue a systems orientation in some program contexts. We want to add the point that just thinking about a program in systems terms will force some tough personal decisions.

Suppose an evaluator finds that some program objectives are too easily attainable to constitute a fair test of program effects and, furthermore, determines that program personnel know this is the case. What does the evaluator do in this situation? Suppose he or she encounters resistance and resentment when pointing out these problems? Or, what if evaluation conclusions are dictated to the evaluator?

A systems-oriented evaluator will search to find the source of these constraints on evaluation activities, believing that they are ultimately linked to someone's values. The evaluator would have to decide how far up (or down) he or she wanted or needed to push. Having moved up or down, the evaluator would then have to decide what to do with the information gained from that activity. Finally, the evaluator must decide what information to share with what program personnel. Is it all private? Would it be beneficial to make some or all of it public? The evaluator needs to be mindful that "An open society becomes a closed society when only the officials know what is going on [Cronbach et al., 1980, p. 95]." These are but a few of the dilemmas facing the systems-oriented evaluator.

Few guidelines exist to assist the evaluator in such dilemmas. Some useful suggestions have been offered by Cronbach et al.:

1. At the outset, in the decision to accept a commission, the evaluator's values should loom large. In particular, we urge that he not attempt to evaluate a program with whose basic aims he is not in sympathy; a pacifist agreeing to conduct an impartial evaluation of ROTC would find himself torn by conflict after conflict. (That the evaluator will be too sympathetic is not a grave danger; the more he believes that an effective program is socially desirable, the keener his search for possible improvements in it will be.)
2. In collecting data, the evaluator strives for openness and neutrality, shaping his techniques to bring both good and bad news to light. When it comes to interpretation, he tries to be as impartial as possible, not by striving for a value-neutral report but by considering the facts from the relevant, no doubt conflicting, value perspectives. He does press to get the possible value-laden perspectives before all the parties, except as agreements he has made proscribe such active communication.
3. Having done his professional job, he puts off his professional robes and, if he chooses, speaks up for what he as a citizen favors [pp. 208-209].

More specifically, these writers make the following suggestions:

1. The evaluator should have an explicit written understanding with the sponsor as to their respective rights to release information.
2. Having entered into such an agreement, the evaluator should abide by it, short of tacit complicity in perjury.
3. The evaluator should not undertake to serve an agency unless he is in sympathy with its general mission.
4. In planning and interpreting his work, the evaluator should think in terms of the larger public interest even when that is inconsistent with the parochial interests of the agency.
5. The evaluator should not agree to let an agency alter or withhold his reports if he has reason to believe that in past instances the agency has distorted evaluative information in releasing it.
6. Provision should be made for full scrutiny, by a panel of qualified reviewers who are not aligned with the sponsoring agency, of any politically controversial finding. Such review may be carried out in camera, when public disclosure would not be in the public interest [pp. 210-211].

Planning and the Evaluative Enterprise

The concepts of *backward* and *forward evaluation* discussed in the previous chapter illustrate our belief that planning and evaluation begin and end with values. In the previous chapter we took special note that values constitute the "stuff" from which programs are planned and with which they should be evaluated. There and elsewhere in this book we have stated the belief that the process of explicating value-based criteria in the program environment, determining the relationship of values to stated program objectives, and using these values to determine the worth and merit of a program are responsibilities the field of evaluation should assume.

This perspective and the responsibilities it implies for the evaluator represent a significant departure from present-day practices in program evaluation. And, as we have noted, this perspective will not always be tolerated or be in the best interests of the evaluator. Yet, it is our conviction that identifying value-based criteria operating in the program environment, even when this is done in the privacy of one's own thoughts, accrues valuable insights and benefits. We believe the fruits of this type of thinking will work their way into richer conceptualizations of a program. Even if this type of thinking only aided in making the evaluator and decision maker more aware of the criteria with which the program was *not* evaluated, this, in our opinion, would substantially improve the credibility and honesty of the evaluation. Placing the evaluation in some larger value framework, whether or not it is actually used in evaluating the program, constitutes a much needed addition to present-day evaluation practice.

A further point we would like to make is that, although planning and evaluation are integrally tied to the same set of values, these values are themselves tied to other sources of data in the program environment. Some of these data have been identified in this volume and have been noted to include the needs of program participants, the knowledge held by technical experts and vested interest groups, qualitative knowing acquired through experience with the program, and data about the conditions or circumstances surrounding the program that could regulate, modify, or constrain its effectiveness. With some reflection it becomes obvious that these sources of data—values, needs, experience, and conditions—all are used routinely by any intelligent person in making judgments and decisions about almost anything. Hence, it should be no surprise that we are suggesting that this larger informational network also be used in the planning and evaluation of programs.

A final point we would like to make concerning the relationship of planning to evaluation is that, although seldom noticed, planning and evaluating are often simultaneously occurring processes. So closely related are these processes in making everyday judgments and decisions that we often are unaware of the point at which our planning ends and an evaluation of the consequences of our plan begins. The flow of information from planning to evaluation and back again is spontaneous and continuous, affording us the opportunity to continuously refine and improve what might initially be only a rough approximation of what is ultimately wanted, needed, and desired. This naturally occurring process both informs *and* forms our values, helps us to assess the conditions under which our plans must operate, and sends back to us a mental image of planned events before we have to incur the responsibilities and consequences inherent in them.

We believe that the planned change process should be conceptualized with a similar lack of distinction between planning and evaluation. We have attempted to show that values and the overarching purpose of planned change do far more to unify planning and evaluation than the specialized tools and techniques within these fields do to separate them. Once the planned change process is understood, it becomes difficult to justify the separation of program planning from program evaluation, just as it would be difficult to separate planning from evaluation in everyday decision making. The point by now has been well made: the activities of planning and evaluation are highly interrelated components of the planned change process. By tying a program (and its evaluation) to the values that undergird it, we establish the means with which to cast both planning and evaluation into a single, unitary framework.

The Systems Perspective within the Context of Emerging Trends in Program Evaluation

Several major trends and developments in the recent history of program evaluation and organizational development are consistent with the perspective we have suggested. Several of these warrant some mention.

One issue receiving considerable attention of late is the distinction between qualitative and quantitative modes of research and the relevance of both for program evaluation. Several theoretical positions can be identified relevant to this issue. Britan (1978) and Filstead (1979), for example, provide recent arguments about the limitations of statistically oriented evaluation models. In pointing out these limitations, Britan suggests that we attempt to "understand multidimensional causes and effects through holistic analysis of program context [p. 230]." Several writers, including Britan, Campbell (1978, 1979b), and Reichardt and Cook (1979) have called for a meshing of qualitative and quantitative modes of inquiry. The general thrust of this position is that distinctions between qualitative and quantitative modes of inquiry are unimportant and waste effort. Implicit is the need for a larger conceptual framework for judging procedural alternatives in program evaluation.

Related to this qualitative–quantitative dimension are some issues we raised in Chapter 5. We argued there that systems notions and concepts have not been readily assimilated in program evaluation practice because of the impression many have that a systems approach is overly complex with respect to the phenomenon being studied and, therefore, cumbersome both conceptually and procedurally. Campbell (1978) points out, however, that

> Systems analysis is a frequently recommended alternative/supplement to the quantitative experimental approach. While in biology and engineering, this is a highly quantified utilization of experimental findings, it has come to represent in social programme evaluation a general qualitative approach typified by an attention to organizational functioning, relationships to other systems, and close attention to the actual processes involved in the treatment [p. 197].

Broader conceptual frameworks for choosing evaluation strategies are appearing in the evaluation literature. Morell (1979), for example, has developed a matrix indicating primary themes of different kinds of outcome evaluation as a function of various information needs. While his conceptualization of these issues are helpful, he considers only outcome evaluation. A broader perspective has been offered by Perkins (1977). He advocated a "contingency approach" to selecting evaluation procedures. He identified six major evaluation types (strategic evaluations, compliance evaluations, program design, management evaluations, evaluations of interventions, and evaluation of program impact) each of which could be addressed by any of nine primary evaluation strategies (e.g., experiment, quasi-experiment, cost–benefit analysis, administrative audits, operations research, etc.). His contingency approach involves a comparative analysis of methodological alternatives as a means of "achieving an optimal fit between evaluation objective and assessment method [p. 652]."

Perkins' major contribution is that he has offered a broad overarching framework for conceptualizing evaluation activity; implicit in his contingency approach is the need for a high degree of eclecticism in evaluation practice. He does not, however, offer much detail about what methodologies might be predisposed to the information needs of certain types of evaluation. Although Perkins provides a rich conceptual matrix for thinking about evaluation, the cells are left empty. They must be "filled" according to the information demands of the specific context.

Another conceptualization of program evaluation activities is depicted in Figure 9.1. This figure is based on Britan's (1978) tripartite view of program evaluation, with some extensions and modifications. We have risked filling in the cells with some educated guesses beyond those provided by Britan. The three dimensions of this figure (theory, goals, and results) are consistent with our ideas about the legitimacy, appropriateness, and representativeness of program objectives and it is offered as a conceptual scheme or heuristic to help one think about which evaluation approaches may be appropriate to

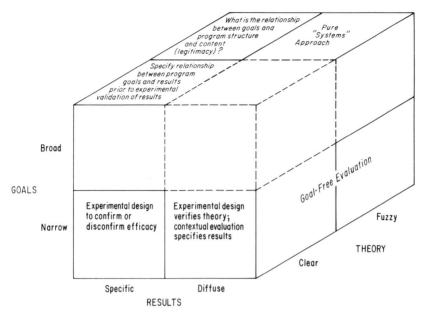

Figure 9.1. **Britan's (1978) view of program evaluation with extensions and modifications.**

specific program contexts. It would seem that the "contingency approach" suggested by Perkins (1977) would have to address some basic program dimensions similar to those depicted in Figure 9.1.

Some aspects of Figure 9.1 are obvious. For example, the cell depicting clear theory, narrow goals, and specific results is analogous to the conditions of the controlled laboratory experiment and it is here that the statistical procedures developed for those situations are most appropriate. Britan (1978) argued that it is difficult to imagine a program based on clear theory having broad goals and diffuse results. Thus the broken lines for this cell. The other cells in the figure are less ambiguous, and suggest methodological alternatives. This figure does place evaluation methodology in a larger perspective, especially the practice of experimental design within the broad context of evaluation practice.

Another conceptual schema for choosing evaluation alternatives was presented in Chapter 5 (Figure 5.1, p. 135). In this framework, complexity was the central issue. That figure illustrated that the capacity for complexity of a solution strategy needs to match the amount of complexity inherent in a problem. For example, the criticisms of traditional experimental design, which we have mentioned throughout this book are based primarily on the notion that such procedures greatly oversimplify the way the world really is.

In other words, the complexity inherent in the problem greatly exceeds the capacity of the solution strategy.

At the other end of the continuum are the early applications of systems analysis to social problems. We reviewed criticisms of these efforts in Chapter 5. The thrust of those criticisms were that the overly quantitative, highly complex approach inherent in systems analysis far outstripped our understanding of and ability to measure the variables of interest in the areas being studied. Although there was plenty of complexity inherent in the problem, our limited understanding of that complexity dampened the usefulness of many of the operations research procedures for studying those systems. Solution strategies with a large complexity-handling capacity should not be applied to simple problems or to problems where our level of understanding does not allow such an elaborate analysis.

From this brief review and discussion it should be evident that conceptualizing evaluation within some larger framework forces one out of a parochial allegiance to specific methodological practices toward a more eclectic stance. The appropriateness of a particular evaluation practice for some specific program context is dependent on a number of dimensions. Making decisions about what would be the best strategy to follow requires *at least* an implicit choice of what dimensions are most salient.

Another trend we see developing on the horizon for program evaluation is the issue of uncertainty. More and more, uncertainty is being permitted a "place" in the way we view programs and program evaluation. The traditionally used cause-and-effect models underlying most statistical procedures allow very little uncertainty. This perspective even attempts to specify the amount of uncertainty that will be "allowed" or tolerated (usually as an arbitrarily selected probability that an event could be expected to occur by chance alone). Many prominent writers and theorists have recently suggested that things are not as "certain" as our statistical models would lead us to assume.

Glass (1979), for example, has suggested that:

> Perhaps today's education should be designed as a system would be designed around any phenomenon that is unlawful, unpredictable, and unknowable in the traditional scientific sense. On the basis of my experience in synthesizing research, I believe that we cannot now explain any appreciable proportion of the variance in success of different methods of schooling implemented in different places. . . . It is a perfectly intellectually respectable endeavor, though you may not think of it as Science, to design a system that will function well under conditions of high uncertainty. In a "general systems theory" sense, we can distinguish among systems with regard to their predictability. From where we stand at this moment, education is a highly unpredictable one. But it certainly does not follow that there exists no rational way to cope with the teaching–learning system simply because it is unpredictable. Take two other systems as analogies: the meteorolog-

ical system and the worldwide oxidation system. In spite of what the weather reports many claim, the meteorological system is highly unpredictable yet we have invented many devices for coping with unpredictable weather: portable umbrellas and windshield wipers, as well as the bucolic invention of the American Midwest, the storm celler.

Some general principles are recognizable in how we attempt to deal with unpredictable systems. Such systems must be monitored diligently; the actors within them must remain versatile and flexible, and the services must be highly decentralized. Persons must command options instead of eternal truths. In education, this style of coping would be the very antithesis of the style of top-down prescriptive planning and policy that we researchers imagine ourselves discovering [p. 14].

Weick (1980) has recently addressed uncertainty in organizations, especially educational organizations, although his comments are equally applicable to other systems, as well. He talks of how "tightly" or "loosely" systems are coupled (in much the same sense as Ashby, 1956, used the term), and how coupling relates to the predictability of individuals and events in organizations. In a tightly coupled system, statements of the if A, then B variety predominate. Here, there is little uncertainty. Loosely coupled systems, Weick argues, are characterized by the statement "if A, then B . . ." followed by any one of seven adverbs: occasionally, eventually, negligibly, gradually, suddenly, ceremoniously, or indirectly.

Although his position is different from that of Glass, his view of uncertainty is not.

Notice that the concept of loose coupling tells you why you can't predict what will happen in many organizations. In this sense the concept has some kinship with meteorology. The science of meteorology explains conditions under which it is impossible to forecast the weather. As we gain a better understanding of weather we understand more clearly why accurate forecasting is impossible. Understanding does not necessarily lead to more accurate predictions. Rather it leads to better predictions about those times when predicting won't work. Loose coupling operates the same way. To understand more about a loosely coupled system is to understand more clearly why predictions about that system will fail. To talk about a loosely coupled system is not to talk about structural looseness but about process looseness. The image is that of a process that unfolds unevenly, discontinuously, sporadically, or unpredictably, if it unfolds at all [Weick, 1980, p. 3].

Weick's description of loosely coupled processes is similar to Cronbach's description of the processes of accommodation and negotiation by which decisions emerge within the policy-shaping community. In agreement with Cronbach et al. and with Weick, we see evaluation as usually occurring in a very loosely coupled context. Thus, the assumption of a context of command, which has characterized traditional decision-oriented evaluation practice, may not always be applicable. Command implies tight coupling; a clearly definable decision maker is believed to issue orders which are followed precisely to bring out a desired result. If the context in which a program actually occurs is actually very loosely coupled, how relevant will the evaluative data, results, conclusions, and recommendations be if the evaluator has assumed all along that he or she was operating within a context of command?

The points made by Weick and Glass are consistent with our own systems view and illustrate the need for a broader conceptual framework for dealing with uncertainty. Taken together these ideas represent attempts to deal with complexity and ambiguity in as direct a manner as possible. These and similar ideas operate as heuristics that can help one be smarter about programs.

Heuristics to help one deal with the "unknowable" are not limited to broad-scale perspectives such as the ones we have elucidated in this book. Recent methodological developments for dealing with ambiguity and lack of consensus have resulted in specific procedures for attacking such problems. These include the Delphi technique (Helmer, 1967), the decision-theoretic approach (Edwards, Guttentag, and Snapper, 1975), and the nominal groups technique (Delbecq, Van de Ven, and Gustafson, 1975). Sjoberg (1975) and Cohen and March (1974) have suggested unconventional thinking styles or patterns if a problem isn't amenable to "treatment" through the usual logical-deductive style that characterizes much of the way we think about the world. Their suggestions are to engage in "playful thinking" (Cohen and March, 1974) and "countersystem analysis" (Sjoberg, 1975). These are mental heuristics as opposed to specific procedures for dealing with uncertainty. The writings of these and many other authors are evidence that alternative approaches to difficult and even intractable problems are evolving.

Summary

The points we have made in this last chapter have not "tied down all the loose ends" so much as they have raised additional questions and issues. This perhaps was intentional and certainly is not inconsistent with the way we have approached other chapters in this book. We have reached a point where we have to put a capstone to this effort. Because this book relates our developing thoughts on the topic of programs and systems and because that train of thought is ongoing rather than finished, it is difficult to end. This chapter reflects, more than any other chapter, the directions of our current thinking.

In this book we have tried to demonstrate that traditional evaluation practice has limited the development of the field and contributed to some of the weaknesses often attributed to program evaluation. We offered a systems perspective as an alternative that might permit the field to grow and develop in a less restricted fashion.

We have tried to promote the idea that a systems view of programs makes one a better evaluator. We believe this perspective can serve as a general heuristic that makes one smarter about programs. A systems thinker is more comfortable with loosely coupled systems, circumstances, and occurrences which happen under conditions of uncertainty. Our experience has been that evaluators are generally asked to bring their expertise to bear in just such vague, ambiguous, loosely coupled program contexts.

Finally, we have tried to appeal to common sense. It is clearly our view that common sense, when applied to high degrees of complexity, ambiguity, and uncertainty, reduces to an understanding of decisions, values, and systems.

We hope we have fostered a clear perception in the minds of our readers that the complexity that overwhelms us and our methodology can be rendered more manageable and less disruptive if we admit, even welcome, its presence as an "essential, nonignorable property" in program evaluation. To accept the confusion inherent in pursuing program evaluation goals is to be enlightened about that confusion. We have been as specific as we can be at this time about how one might "manage" confusion through a systems perspective. Admitting to chaos, complexity, and uncertainty is generally avoided at all costs for it is thought to engender excessive, debilitating anxiety. Our experience suggests just the opposite: admitting that this is present and ongoing can be quite relaxing, even reassuring. Why? Because a systems perception of the real world matches the variety found within it. We suggest this is a small price to pay for methodological peace of mind. Rather than avoiding or being overwhelmed by complexity, we hope the program planner and evaluator will have the common sense to accept it and the curiosity to understand it.

APPENDIX:

An Introduction to Program Modeling[1]

Overview

PROGRAM MODELING

Program modeling is a graphic methodology for describing, analyzing and communicating program planning, development, and evaluation concepts. Program models can assist administrators, planners, developers, and evaluators in clarifying program objectives, identifying program activities, and specifying program outcomes. Program modeling accomplishes this by forcing the articulation of program components and their interrelationships in clear and precise notation.

Program models can be used for:

1. Thinking in a structured way about a complex program and getting stakeholders to react to its meaning.
2. Communicating planning and design concepts to legislators, program developers, evaluators, administrators, and program participants
3. Insuring the accuracy, completeness, and quality of an evolving program description
4. Managing and guiding the development of a project
5. Identifying standards and criteria for judging the effectiveness of a program

[1]The contents of this appendix are illustrated with an adaptation of the Structured Analysis and Design Technique (SADT). SADT was developed as a general modeling tool by Doug Ross (Ross, 1977; Ross and Schoman, 1977). This methodology has been adapted to planning and evaluation in the social and behavioral sciences by Gary Borich (Borich, 1979; Borich and Brackett, 1978).

241

STRUCTURED DECOMPOSITION

Structured decomposition is the process of breaking down a program into its component parts. Decomposition begins by graphically dividing the program into a number of general activities, each symbolizing a major class of events within the program. Each general activity is then further broken down into a small number of subactivities in succeeding steps of the process. By introducing substantive detail gradually, and in meaningful steps, a uniform, systematic exposition of successive levels of detail is achieved at the same time that a global conceptualization of the program is maintained (see Figure A.1).

Understanding of a program or system increases by crossing strata; in moving down the hierarchy one obtains a more detailed explanation, whereas in moving up the hierarchy, one obtains a deeper understanding of its significance. In moving up the hierarchy one finds broader concepts and longer periods of time, thus allowing the meaning of the program to be interpreted in more general terms. In moving down the hierarchy one finds narrower concepts and shorter periods of time, allowing for a more precise and detailed explanation of how the program functions to carry out its objectives.

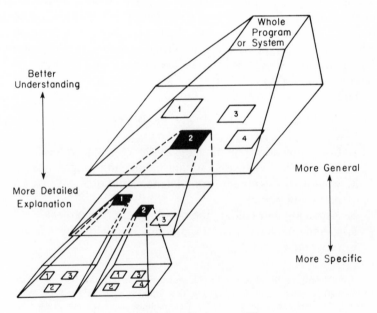

Figure A.1. **A model indicating the decomposition process.**

There are several common features of a program model:

1. A higher-level unit is concerned with a larger portion or broader aspects of the overall system behavior.
2. The decision period at a higher-level unit is longer than that at lower units.
3. A higher-level unit is concerned with slower, more time-consuming aspects of the overall system behavior.
4. Descriptions and problems at higher levels are less structured, with more uncertainties, and are more difficult to formalize quantitatively.

Basic Concepts and Designations

TRANSACTIONS

Transactions are the basic units or building blocks of a program model. A transaction defines a well-constrained context for specifying further levels of detail. In modeling a program, transactions are indicated as boxes and labeled to indicate their purpose. Transactions are program activities and thus should be labeled with an active verb (see Figure A.2).

```
┌─────────────┐
│   Teach     │
│   Word      │
│ Recognition │
└─────────────┘
                    ┌─────────────┐
                    │   Present   │
                    │ Vocabulary  │
                    │ in Sentences│
                    └─────────────┘
                                    ┌─────────────┐
                                    │   Provide   │
                                    │  Readings   │
                                    │    with     │
                                    │ Supervision │
                                    └─────────────┘
```

Figure A.2. **Some examples of transactions.**

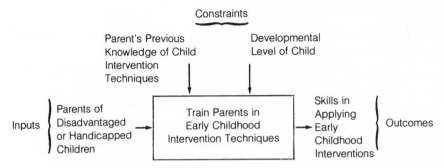

Figure A.3. **Designations used to indicate inputs, constraints, and outcomes.**

INPUTS, CONSTRAINTS, AND OUTCOMES

Inputs, constraints, and outcomes clarify the activity implied by the transaction and show its relationship to other transactions. Transactions are defined by who is being trained or provided the service (inputs), the variables that can affect the training or provision of service (constraints), and the behaviors or conditions that are expected to result (outcomes). See Figure A.3. To distinguish their separate functions, inputs are shown as arrows entering the left side of the transaction, constraints as arrows entering the top of the transaction, and outcomes as arrows emerging from the right side of the transaction.

Inputs: Inputs are things the activity uses or acts on; clients or students, tests, staff, time, classrooms, etc.
Constraints: Constraints are things that constrain the activity; funding, program priorities, feedback on results, previous behaviors of the participants, anything that moderates or influences outcomes.
Outcomes: Outcomes are things "produced" or resulting from the activity; clients with improved outlook, higher rates of employment, students with a certain skill.

INTERRELATIONSHIPS AMONG INPUTS, CONSTRAINTS,
AND OUTCOMES

Input, constraint, and outcome designations reveal how transactions within a program are interrelated (see Figure A.4).

THE MEANS-END CONTINUUM

The concept of means–end refers to the notion that any event or condition (e.g., a clinical intervention, a particular teaching strategy, a health care

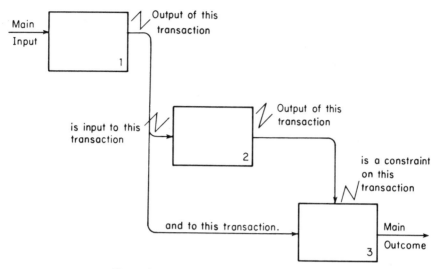

Figure A.4. **Relationships among transactions.**

system, etc.) can be viewed as occupying space on a continuum such that it is simultaneously an end to those events and conditions that preceded it and a means to those that follow. Although each event, condition, or program has its own "ends in view" these ends should be congruent with higher-level goals to which they are expected to contribute. By introducing substantive detail one step at a time, and indicating the interrelationships among program activities, the program model reveals a picture of how program activi-

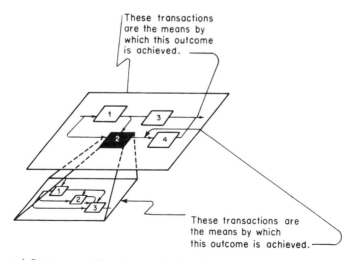

Figure A.5. **Decomposition diagram showing the means–end continuum for a program.**

ties at one level of detail become the means by which outcomes at a higher level are achieved (see Figure A.5).

Beginning the Decomposition Process

STEP 1

Decomposition begins with a phrase that concisely captures the purpose of the whole program. This phrase bounds the context for all subsequent levels of detail and is shown as a transaction at the top—or most general level—of the program model.

STEP 2

From this description, the program is decomposed into two to six clusters or lumpings of activities each expressed at approximately the same level of generality. These general clusters are shown as transactions arranged diagonally and in the sequence in which they occur in the program. See Figure A.6. If sequence is not relevant, the order of transactions is arbitrary.

STEP 3

Each of the clusters identified in Step 2 is further decomposed in separate diagrams (see Figure A.7). Transactions continue to be broken down into their functional components until the activity implied by the topmost trans-action is fully clarified or until transactions no longer can be identified for which there are measurable outcomes. Transactions at general levels of the

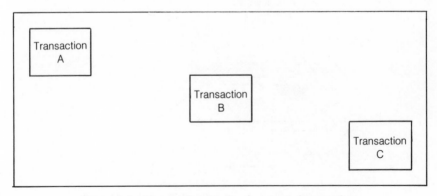

Figure A.6. **Diagram showing the placement of transactions.**

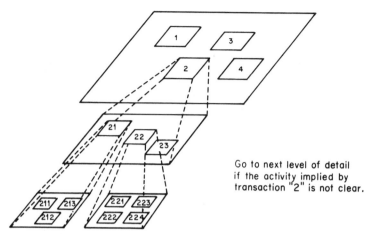

Go to next level of detail
if the activity implied by
transaction "2" is not clear.

Figure A.7. **Diagram showing the decomposition of a general transaction and the interlocking of diagrams.**

program model are expected to account for larger portions of the variance in terminal outcomes and span more time than transactions at detailed levels of the model. This is because as the program becomes more decomposed, transactions at the lower, more detailed levels of the model represent smaller portions of the whole. Successive diagrams within a model are interlocked with a number placed in the lower right corner of each transaction. This number uniquely identifies each transaction's place in the larger program or system.

In deciding on the appropriate level of detail at which to model a particular program or system, the following factors should be taken into consideration.

Time available. The amount of detail shown by the decomposition model should reflect the time allotted the evaluator for conducting the evaluation. It is estimated that decomposing a program can consume as much as a third of the time an evaluator might ordinarily devote to an evaluation project. Obviously the amount of detail that could be shown in different types of evaluations might be vastly different, but presumably so would the complexity of the program to be modeled and the generality of conclusions expected of the evaluator by clients and program management.

Level of detail at which the program was designed. The decomposition model should reflect the level of detail at which the program was designed. Forcing levels of detail for which there is no substantiation generally only increases the inaccuracy of the model, requiring revision of the model at a later time. Any detailing of the program beyond the level at which the program was conceptualized would constitute program design.

On the other hand, sometimes clues to a program's structure can be picked up informally when the program's structure has not been formally described in program documents or by program staff. In that all aspects of the decomposition model should be submitted to program staff for comment and review, first draft diagrams may legitimately constitute hypotheses as to how the program is structured, if there are indications of this structure from informal but observable data. In this manner the decomposition model can serve as a vehicle by which program staff can be questioned about the program structure and content and be alerted to the need for more substance and detail where this may be needed.

Level of generality at which decisions about the program will be made. It is the frame of reference for making decisions about a program that should indicate the level of specificity to be communicated by the decomposition model. When decision making focuses on specific transactions, the program model should reflect a greater level of detail than when decisions focus on higher order clusters of transactions. This level of decision specificity should be determined before program modeling begins.

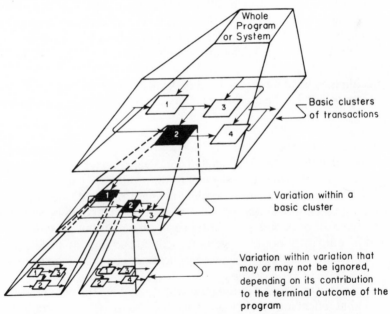

Figure A.8. **Model showing three levels of variation and meaningfulness to terminal outcome. Depending on the program, more or fewer levels of variation may be desirable.**

Meaningfulness of specific transactions to the overall purpose of the program. When decomposing general transactions into more specific transactions, it must be decided whether these more specific transactions contribute in meaningful ways to the terminal outcome of the program. "Splitting hairs" at the lower levels of the decomposition model may introduce irrelevant detail and may even hide the importance of higher-level transactions. Overmodeling uses up valuable time and, if actually incorporated into the evaluation, could prevent more meaningful activities from being completed.

To avoid overmodeling three types of variation within a program can be determined: (a) variation expressed as basic clusters or lumpings of program activities from which all further detail originates, (b) variation contained within the basic clusters that must be recognized in order for the model to comprehensively and accurately reflect the program, and (c) variations in the activities contained within these (higher-level) variations that should be ignored in order not to model miniscule events that do not meaningfully contribute to the terminal outcome of the program (see Figure A.8).

Measurable outcomes. If one or more of the above considerations does not terminate the modeling process, the decomposition should end when program activities can no longer be identified for which there are measurable outcomes. Transactions at which it has been decided to end the modeling process should be decomposed one further level in order to check (a) if that level is the last for which measurable outcomes can be identified and (b) if the transactions at which the model is to end actually represent units of program activity that have meaningful functions (and not simply empty labels). Even when further decomposition is not needed, the process of decomposing final level transactions in the same manner as higher-level transactions generally leads to a fuller understanding of the last level of usable detail.

Conventions of Program Modeling

The following steps apply to each diagram in the decomposition model after all transactions have been decomposed and arranged on diagrams in a descending order of generality.

Step 1. Determine the Main Input and Output; Draw and Label These on the Diagram.

The main input identifies who is to be provided the transactions and the main output identifies the behavior or condition the participants are expected to attain from these transactions. Main outputs can describe be-

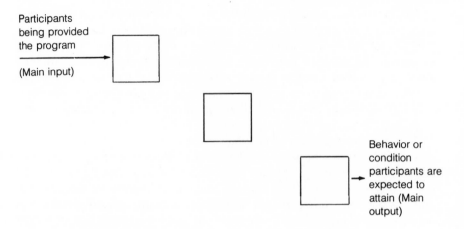

Figure A.9. **Diagram showing main input and output.**

haviors or conditions generally at higher levels of the model but should describe behaviors or conditions specifically at lower levels of the model. Outcomes should be expressed in standard behavioral terms consistent with the terminology used by program staff (see Figure A.9).

Step 2. Draw the Main Path Indicating the Relationships Among Transactions.

Interrelationships may or may not show sequence, as the following alternatives note.

A. *Simple sequence.* What is attained from A affects outcome of B, what is attained from B affects outcome of C, etc. Transactions must be taken in the sequence shown in Figure A.10.

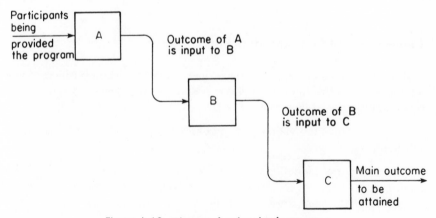

Figure A.10. **Diagram showing simple sequence.**

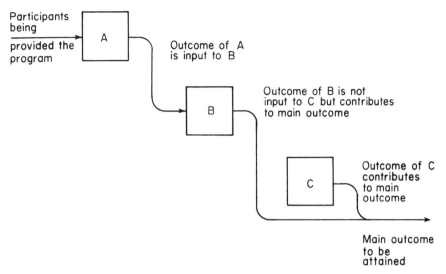

Figure A.11. **Diagram showing mixed sequence.**

B. *Mixed sequence.* What is attained from A effects outcome of B but outcome of B does not affect outcome of C. A and B need not always precede C but A always precedes B (see Figure A.11).

C. *No sequence.* All transactions are independent of one another. They may be taken in any order. (When inputs are not shown they can be assumed to be the program participants. See Figure A.12.)

Step 3. Provide the Constraints that Are Likely to Influence Implementation of Each Transaction or Its Outcome.

Constraints may comprise aptitudes, attitudes, and achievements of trainees and trainers or contextual factors, such as policies, regulations, and funding, which may either hinder or aid implementation of the transaction or the outcomes participants are expected to attain from it. Label each constraint consistent with the level of detail at which the transaction is described, as in Figure A.13.

Step 4. Add Inputs and Outputs that Are Not on the Main Path.

Other Inputs. Specify material, equipment, human resources, or facilities (e.g., workbook, technical manuals, guest lecturer, special classroom) whose presence is considered critical to implementing the transaction. These inputs generally comprise materials, equipment, human resources, or facilities that are so integrally tied to the transaction that it cannot function without them. Including these inputs reminds program developers, evaluators, and program personnel of the importance of the designated items

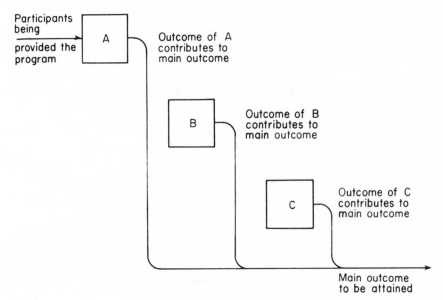

Figure A.12. **Diagram showing no sequence.**

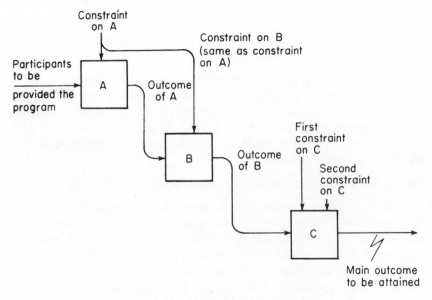

Figure A.13. **Diagram showing constraints.**

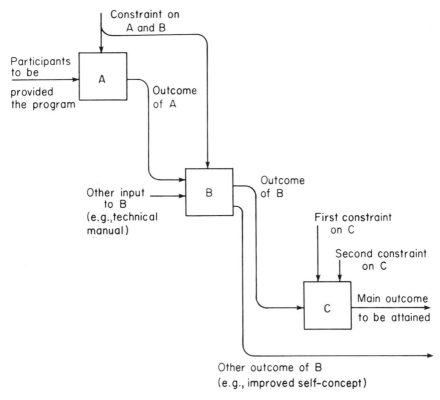

Figure A.14. **Diagram showing other inputs and outcomes.**

to the successful functioning of the transaction. Other inputs serve the same function as constraints but differ from aptitudes, achievements, and contextual factors in that they usually are evaluated on the basis of their presence or absence.

Other outcomes. Other outcomes identify secondary results and side effects that do not directly contribute to the main outcome of the diagram. Changes in attitude, self-concept, physical stamina, or study habits, whether planned or unplanned, represent varieties of other outcomes. See Figure A.14.

Some Additional Conventions Regarding Constraint and Sequence

The following are some varieties of constraint and sequence and their implications for program modeling.

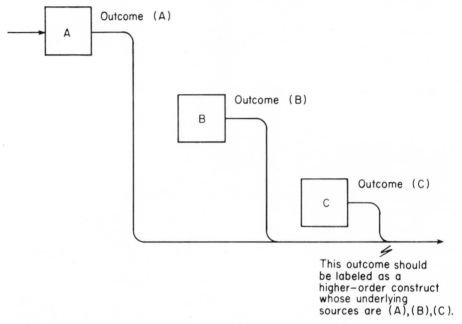

This outcome should be labeled as a higher-order construct whose underlying sources are (A),(B),(C).

Figure A.15. **Diagram showing the independent development of a higher-order construct from lower-order constructs. This higher-order construct is independent of sequence.**

1. A diagram on which no sequence is shown results in a main outcome different than the outcome of any of its transactions (see Figure A.15).

2. A diagram showing sequence can result in a main outcome that is the same as its last transaction, as shown in Figure A.16.

3. Inputs, constraints, and outputs for a parent transaction should be consistent with inputs, constraints, and outcomes on the detail diagram. In large models with many interconnecting arrows, it is desirable to interlock inputs, constraints, and outcomes across diagrams with numerical codes. These codes comprise a letter indicating the origin of an arrow as input (I), constraint (C), or outcome (O) and a number indicating its *ordinal position on the parent diagram* (see Figure A.17).

4. Whereas all inputs, constraints, and outcomes must be consistent across levels of detail, inputs, constraints, and outcomes at higher levels may be expressed more generally than their counterparts at lower levels. This convention is used when an input, constraint, or outcome at a higher level represents related but different behaviors that are to be deployed differently at lower levels. Arrows at higher levels of the model are similar to coaxial cables or pipelines carrying conceptually related but different variables. The internal contents of these "pipes" or "cables" can be peeled away at lower levels as needed (see Figure A.18).

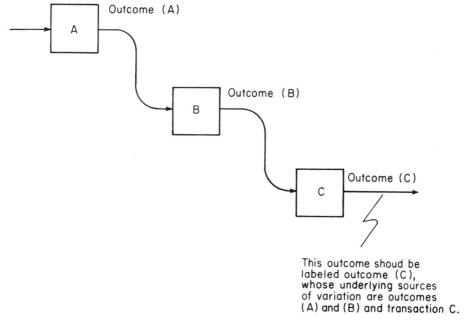

Figure A.16. **Diagram showing the successive development of a higher-order construct from lower-order constructs. This higher-order construct is attained only by sequence.**

5. Feedback enables a program to be self-regulating or to adapt to its own environment. A feedback loop is a mechanism that senses the preliminary output of a transaction and steers the subsequent action of another transaction so that eventually the desired result or "target behavior" is obtained. At least two different types of feedback can be modeled:

A. Feedback in which the program participant recycles through a portion of the program. This type of feedback occurs when a participant "retakes" a part of the program in order to attain a competency level not reached the first time through. This type of feedback is shown with an *output to input* arrow in Figure A.19.

B. A second type of feedback is that in which information about the performance of program participants is sent to some earlier portion of the program for the purpose of revising the nature of the program given to subsequent program participants. For example, if performance levels of program participants fail to reach the criterion expected at program completion, this information may be sent to initial program components to provide the basis for revising the program for the next group of participants. This type of feedback is shown with an *output to constraint* arrow in Figure A.20.

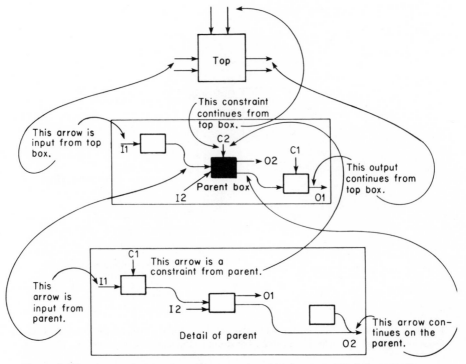

Figure A.17. **Diagram showing the coding system by which inputs, constraints, and outcomes are interlocked across levels of detail.**

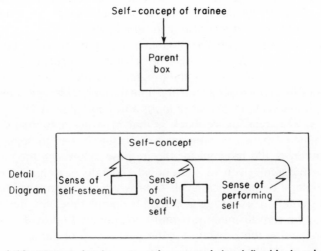

Figure A.18. **Diagram showing a general construct being defined by its subconstructs.**

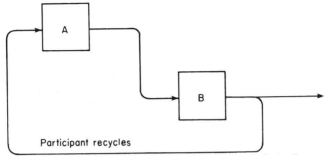

Figure A.19. **Participants who fail to attain the specified behavior condition at transaction B return to transaction A.**

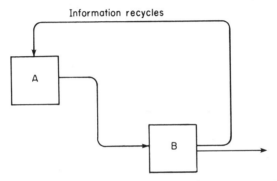

Figure A.20. **Information about the performance of participants at transaction B is used to revise the program at transaction A.**

Checking the Completed Model

The following are some suggestions for checking the accuracy and thoroughness of the completed model.

1. Can each transaction actually be observed? Does each represent an observable activity as opposed to a covert, introspective, or mental activity whose implementation may be difficult or impossible to measure?
2. Are outcomes at the lower levels of the model stated behaviorally and can operational definitions be expressed for them?
3. Are all important constraints specified? Can any less important ones be eliminated to simplify the message being conveyed by the diagram? To assure the importance of each constraint, ask how it helps or hinders the participants attain the outcome intended.

4. Do the labels affixed to transactions and outcomes express a consistent viewpoint? Generally, transactions should indicate the instruction or service that is to be provided, and outcomes the behaviors or conditions that result. To assure a consistent viewpoint check to determine if all transactions are stated from a single point of view.

5. Are there logical mismatches between transactions and outcomes? That is, is the transaction either more or less comprehensive than the outcome specified for it? Is the activity implied by the transaction too little to produce the outcome expected? Or is the activity more than what is needed to be efficient and cost-effective in producing the outcome? This could be a design flaw in the program and not a modeling error.

6. Do there appear to be insufficient or missing transactions or outcomes? Are there holes or gaps in the model where added instruction or services are needed as bridging activities before terminal outcomes can be attained? This also could be a design flaw and not a modeling error.

References

Adams, S. *Evaluative research in corrections: A practical guide.* U.S. Department of Justice, Washington, D.C.: U.S. Government Printing Office, March 1975.

Alkin, M. C. Evaluation theory development. *Evaluation Comment,* 1969, *2*(1), 2-7.

Anderson, S. B., and Ball, S. *The profession and practice of program evaluation.* San Francisco, Calif.: Jossey-Bass, 1978.

Apple, M. W. The process and ideology of valuing in educational settings. In M. Apple, M. Subkoviak, and H. Lufler, Jr. (Eds.), *Educational evaluation: Analysis and responsibility.* Berkeley, Calif.: McCutchan, 1974.

Ashburn, A. G. Credibility gaps and the institutionalizing of educational evaluation functions. *Planning and Changing: A Journal for School Administration,* 1973, *4*(1), 18-28.

Ashby, W. R. *An introduction to cybernetics.* London: Methuen and Co., 1956.

Astin, A. W., and Panos, R. J. The evaluation of educational programs. In R. L. Thorndike (Ed.), *Educational measurement* (2nd ed.). Washington, D.C.: American Council on Education, 1971.

Attkisson, C. C., and Broskowski, A. Evaluation and the emerging human service concept. In C. C. Attkisson, W. A. Hargreaves, M. J. Horowitz, and J. E. Sorensen (Eds.), *Evaluation of human service programs.* New York: Academic Press, 1978.

Attkisson, C. C., Brown, T. R., and Hargreaves, W. A. Roles and functions of evaluation in human service programs. In C. C. Attkisson, W. A. Hargreaves, M. J. Horowitz, and J. E. Sorensen (Eds.), *Evaluation of human service programs.* New York: Academic Press, 1978.

Attkisson, C. C., Hargreaves, W. A., Horowitz, M. J., and Sorensen, J. E. (Eds.), *Evaluation of human service programs.* New York: Academic Press, 1978.

Ayres, R. U. *Technological forecasting and long range planning.* New York: McGraw-Hill, 1969.

Baker, F. Organizations as open systems. In F. Baker (Ed.), *Organizational systems: General systems approaches to complex organizations.* Homewood, Ill.: Richard D. Irwin, 1973.

Banathy, B. *Instructional systems.* Palo Alto, Calif.: Fearon Publishers, 1968.

Barker, R. G. *Ecological psychology.* Stanford, Calif.: Stanford University Press, 1968.

Barker, R. G. Explorations in ecological psychology. *American Psychologist,* 1965, *20*(1), 1-14.

259

Becker, H. Whose side are we on? In G. Riley (Ed.), *Values, objectivity and the social sciences.* Reading, Mass.: Addison-Wesley, 1974.

Beer, S. *Platform for change.* New York: John Wiley and Sons, 1975.

Beishon, J., and Peters, G. *Systems behavior.* New York: Harper and Row, 1976.

Bell, R., Warheit, G., and Schwab, J. Needs assessment: A strategy for structuring change. In R. D. Coursey, (Ed.), *Program evaluation for mental health: Methods, strategies, participants.* New York: Grune and Stratton, 1977.

Bell, R. A., Nguyen, T. D., Warheit, G. J., and Buhl, J. M. Service utilization, social indicator, and citizen survey approaches to human service need assessment. In C. C. Attkisson, W. A. Hargreaves, M. J. Horowitz, and J. E. Sorensen (Eds.), *Evaluation of human service programs.* New York: Academic Press, 1978.

Bertalanffy, L. An outline of general system theory. *British Journal of Philosophy of Science,* 1950, *1,* 139–164.

Bertalanffy, L. *General systems theory.* New York: Braziller, 1968.

Bessent, A., and Bessent, W. Determining the comparative efficiency of schools through data envelopment analysis. *Educational Administrative Quarterly,* 1980, *16*(2), 57–76.

Binner, P. Goal attainment model versus systems approach. In B. Willer, G. Miller, and L. Cantrell (Eds.), *Information and feedback for evaluation.* Toronto: York University, 1975.

Bogdan, R., and Taylor, S. J. *Introduction to qualitative research methods: A phenomenological approach to the social sciences.* New York: John Wiley and Sons, 1975.

Borich, G. *The appraisal of teaching: Concepts and process.* Reading, Mass.: Addison-Wesley, 1977.

Borich, G. A systems approach to the evaluation of training. In H. F. O'Neil (Ed.), *Procedures for instructional systems development.* New York: Academic Press, 1979.

Borich, G. D. *Evaluating educational programs and products* (Ed.), Englewood Cliffs, N.J.: Educational Technology Publications, 1974.

Borich, G. D., and Brackett, J. Instructional design and evaluation with a structured analysis and design technique. *Educational Technology,* July 1978, 18–23.

Borich, G., and Jemelka, R. Evaluation. In H. F. O'Neil (Ed.), *State of the art assessments in computer-based instruction.* New York: Academic Press, 1981.

Borich, G., and Madden, S. *Evaluating classroom instruction: A sourcebook of instruments.* Reading, Mass.: Addison-Wesley, 1977.

Boruch, R. F., and Gomez, H. Sensitivity, bias and theory in impact evaluations. *Professional Psychology,* 1977, *8*(4), 411–434.

Boulding, K. E. General systems theory—the skeleton of science. *Management Science,* 1956, *2,* 197–208.

Brandl, J. E. Evaluation and politics. *Evaluation,* 1978, Special Issue, 6–7.

Branson, R. K. The interservice procedure for instructional system development. *Educational Technology,* 1978, *18*(3), 11–14.

Braybrooke, D., and Lindblom, C. *A strategy of decision: Policy evaluation as a social process.* New York: The Free Press, 1963.

Brickell, H. M. The influence of external political factors on the role and methodology of evaluation. *Evaluation Comment,* 1976, *5*(2), 2–6.

Briggs, L. (Ed.). *Instructional design.* Englewood Cliffs, N.J.: Educational Technology Publications, 1977.

Britan, G. Experimental and contextual models of program evaluation. *Evaluation and Program Planning,* 1978, *1*(3), 229–234.

Broskowski, A., and Driscoll, J. The organizational context of program evaluation. In C. C. Attkisson, W. A. Hargreaves, M. J. Horowitz, and J. E. Sorensen (Eds.), *Evaluation of human service programs.* New York: Academic Press, 1978.

Broskowski, A., Driscoll, J., and Schulburg, H. C. A management information and planning system for indirect services. In C. C. Attkisson, W. A. Hargreaves, M. J. Horowitz, and J. E. Sorensen. *Evaluation of human service programs.* New York: Academic Press, 1978.

Brown, D. G. A scheme for measuring the output of higher education. In *Seminar on Outputs of Higher Education.* Boulder, Colo.: Western Interstate Commission for Higher Education, July 1970.

Buckley, W. (Ed.), *Modern systems research for the behavioral scientist.* Chicago: Aldine, 1968.

Bunker, D. R. Organizing evaluation to serve the needs of program planners and managers. *Evaluation and Program Planning,* 1978, *1,* 129-234.

Butman, J., and Fletcher, J. The role of the evaluator and developer in educational research and development. In G. Borich (Ed.), *Evaluating educational programs and products.* Englewood Cliffs, N.J.: Educational Technology Publications, 1974.

Campbell, D. T. Qualitative knowing in action research. In M. Brenner, P. Marsh, and M. Brenner (Eds.), *The social contexts of method.* London: Croom Helm, Ltd., 1978.

Campbell, D. T. Assessing the impact of planned social change. *Evaluation and Program Planning,* 1979, *2,* 67-90. (a)

Campbell, D. T. "Degrees of freedom" and the case study. In T. D. Cook and C. S. Reichardt. *Qualitative and quantitative methods in evaluation research.* Beverly Hills, Calif.: Sage Publications, 1979. (b)

Carter, D. E., and Newman, F. L. *A client-oriented system of mental health service delivery and program management: A workbook and guide.* (DHEW Publication No. ADM 76-307). Washington, D.C.: U.S. Government Printing Office, 1976.

Churchman, C. W. *The systems approach.* New York: Delacorte Press, 1968.

Churchman, C. W. *The design of inquiring systems.* New York: McGraw-Hill, 1971.

Churchman, D. A new approach to evaluating the implementation of innovative educational programs. *Educational Technology,* May 1979, 25-28.

Cohen, M. D., and March, J. G. *Leadership and ambiguity: A study of 101 college presidents.* New York: McGraw-Hill, 1974.

Coleman, J. S., and others. *Policy issues and research design.* Chicago: National Opinion Research Center, 1979.

Cook, T. D., and Campbell, D. T. The design and conduct of quasi-experiments and true experiments in field settings. In M. D. Dunnette (Ed.), *Handbook of industrial and organizational psychology.* Chicago: Rand McNally, 1975.

Cook, T. D., and Reichardt, C. S. (Eds.) *Qualitative and quantitative methods in evaluation research.* Beverly Hills, Calif.: Sage Publications, 1979.

Cooley, W. W., and Lohnes, P. R. *Evaluation research in education theory: Principles and practice.* New York: John Wiley and Sons, 1976.

Coursey, R. D. An overview of techniques and models of program evaluation. In R. D. Coursey (Ed.), *Program evaluation for mental health: Methods, strategies, participants.* New York: Grune and Stratton, 1977.

Cronbach, L., Ambron, S., Dornbusch, S., Hess, R., Hornik, R., Phillips, D., Walker, D., and Weiner, S. *Toward reform of program evaluation.* San Francisco: Jossey-Bass, 1980.

Culbertson, J. *Preparing educational leaders for the seventies.* Columbus, Ohio: University Council for Educational Administration, 1969.

Davis, R., Alexander, L., and Yelon, S. *Learning systems design.* New York: McGraw-Hill, 1975.

DeGreene, K. B. Systems and psychology. In J. Beishon and G. Peters (Eds.), *Systems behavior.* New York: Harper and Row, 1976.

Delbecq, A., Van de Ven, A., and Gustafson, D. *Group techniques for program planning.* Glennview, Ill.: Scott, Foresman and Co., 1975.

Demone, H. W., Schulberg, H. C., and Broskowski, A. Evaluation in the context of developments in human services. In C. C. Attkisson, W. A. Hargreaves, M. J. Horowitz, and J. E. Sorensen (Eds.), *Evaluation of human service programs.* New York: Academic Press, 1978.

Deutsch, K. Mechanism, organism and society. *Philosophy of Science,* 1951, *18*(3), 230–252.

Dewey, J. *Human nature and conduct.* New York: Holt, 1922.

Dewey, J. Theory of valuation. *International Encyclopedia of Unified Science,* 1939, *2*(4), 1–67.

Dick, W., and Carey, L. *The systematic design of instruction.* Glenview, Ill.: Scott, Foresman and Co., 1978.

Drucker, P. F. *The practice of management.* New York: Harper and Row, 1954.

Drucker, P. F. *Managing for results.* New York: Harper and Row, 1964.

Drucker, P. F. New templates for today's organizations. *Harvard Business Review,* 1974, *52,* 45–53.

Edwards, W., Guttentag, M., and Snapper, K. A decision-theoretic approach to evaluation research. In E. L. Struening and M. Guttentag (Eds.), *Handbook of evaluation research.* Beverly Hills, Calif.: Sage Publications, 1975.

Eisner, E. W. The perceptive eye: Toward the reformation of educational evaluation. *Occasional Papers of the Stanford Evaluation Consortium.* Stanford, Calif.: Stanford University, 1975.

Eisner, E. W. *The educational imagination.* New York: Macmillan, 1979.

Elpers, J. R., and Chapman, R. L. Basis of the information system design and implementation process. In C. C. Attkisson, W. A. Hargreaves, M. J. Horowitz, and J. E. Sorensen (Eds.), *Evaluation of human service programs.* New York: Academic Press, 1978.

Emery, F. *Systems thinking.* New York: Penguin, 1978.

Emery, F. E. (Ed.) *Systems thinking.* Middlesex, England: Penguin, 1969.

Etzioni, A. Two approaches to organizational analysis: A critique and a suggestion. *Administrative Science Quarterly,* 1960, *5,* 257–278.

Filstead, W. J. Qualitative methods: A needed perspective in evaluation research. In T. D. Cook and C. S. Reichardt (Eds.), *Qualitative and quantitative methods in research.* Beverly Hills, Calif.: Sage Publications, 1979.

Fisher, R. A. *Statistical methods for research workers* (1st ed.). Edinburgh, Scotland: Oliver and Boyd, 1925.

Flaherty, E. W., and Morell, J. A. Evaluation: Manifestations of a new field. *Journal of Evaluation and Program Planning,* 1978, *1,* 1–10.

Forrester, J. W. A new corporate design. *Industrial Management Review,* 1965, *7*(1), 5–17.

Forrester, J. W. *Collected papers of Jay W. Forrester.* Cambridge, Mass.: Wright–Allen Press, 1975.

Friedman, M. *Free to choose* (A 10-part TV series). Erie, Penn.: WQLN Public Communications, Inc. (Producer), 1979.

Friedman, M., and Friedman, E. *Free to choose.* New York: Harcourt Brace Jovanovitch, 1980.

Furst, E. J. *Constructing evaluation instruments.* New York: David McKay, 1964.

Gibson, E. J. *Principles of perception, learning and development.* New York: Appleton-Century-Crofts, 1969.

Glass, G. V. Policy for the unpredictable (uncertainty research and policy). *Educational Researcher,* 1979, *8*(9), 12–14. Copyright 1979, American Educational Research Association, Washington, D.C.

Grobman, H. Evaluation activities of curriculum projects: A starting point. *AERA Monograph Series on Curriculum Evaluation, No. 2: Evaluation Activities of Curriculum Projects.* Chicago: Rand McNally, 1968.

Guba, E. *Toward a methodology of naturalistic inquiry in educational evaluation.* Los Angeles: Center for the Study of Evaluation, University of California, 1978.

Guba, E., and Lincoln Y. S. *Effective Evaluation.* San Francisco: Jossey-Bass, 1981.

Guba, E. G. The failure of educational evaluation. *Educational Technology,* May 1969, 29-38.

Gurel, L. The human side of evaluating human services programs: Problems and prospects. In M. Guttentag and E. L. Struening (Eds.), *Handbook of evaluation research* (Vol. 2). Beverly Hills, Calif.: Sage Publications, 1975.

Hagedorn, H. J., Beck, K. J., Neubert, S. F., and Werlin, S. H. *A working manual of simple program evaluation techniques for community mental health centers.* Rockville, Maryland: National Institute of Mental Health, 1976.

Halpern, J. Program evaluation, systems theory, and output value analysis: A benefit/cost model. In R. D. Coursey (Ed.), *Program evaluation for mental health: Methods, strategies, participants.* New York: Grune and Stratton, 1977.

Halpern, J., and Binner, P. R. A model for an output value analysis of mental health programs. *Administration in Mental Health,* Winter 1972, 40-51.

Hammond, R. Evaluation at the local level. In B. Worthen and J. Sanders, *Educational evaluation: Theory and practice.* Worthington, Ohio: Charles A. Jones, 1973.

Handy, R. *Value theory and the behavioral sciences.* Springfield, Ill.: Charles C Thomas, 1969.

Handy, R., and Kurtz, P. *A current appraisal of the behavioral sciences.* Great Barrington, Mass.: Behavioral Research Council, 1964.

Hardin, G. The cybernetics of competition: A biologist's view of society. *Perspectives in Biology and Medicine,* 1963, 7, 61-84.

Harless, J. H. An analysis of front-end analysis. *Improving Human Performance: A Research Quarterly,* 1973, 2, 229-244.

Haskew, L. D. Force analysis. In S. P. Hencley, and J. R. Yates (Eds.), *Futurism in Education: Methodologies.* Berkeley, Calif.: McCutchan, 1974.

Helmer, O. *Analysis of the future: The delphi technique.* Santa Monica, Calif.: Rand Corp., 1967.

Hemphill, J. The relationships between research and evaluation studies. In R. Tyler (Ed.), *Educational evaluation: New roles, new means, the sixty-eighth yearbook of The National Society of the Study of Education.* Chicago: University of Chicago, 1969.

Hencley, S. P., and Yates, J. R. (Eds.). *Futurism in education: Methodologies.* Berkeley, Calif.: McCutchan, 1974.

Hersh, C. Social history, mental health and community control. *American Psychology,* 1972, 27, 749-454.

Hesse, M. Models versus paradigms in the natural sciences. In L. Collins (Ed.), *Use of models in the social sciences.* London: Tavistock Publications, 1976.

Hinricks, H. H. Government decision-making and the theory of benefit-cost analysis: A primer. In H. H. Henricks and G. M. Taylor (Eds.), *Program budgeting and benefit-cost analysis.* Pacific Palisades, Calif.: Goodyear, 1969.

Hoban, C. Educational technology and human values. *Audiovisual Communications Review,* 1977, 25(1), 221-242.

Hoetker, J., with Fichtenau, R., and Farr, H. L. K. *Systems, systems approaches, and the teacher.* Urbana, Ill.: National Council of Teachers of English, 1972.

Hoos, I. R. *Systems analysis in social policy: A critical review.* London: The Institute of Economic Affairs, 1969.

Horst, P., Nay, J. N., Scanlon, J. W., and Wholey, J. S. Program management and the federal evaluator. *Public Administration Review.* July/August 1974, 33-39.

Horst, D. P., Tallmadge, G. K., and Wood, C. T. *A practical guide to measuring project impact on student achievement.* Washington, D.C.: U.S. Government Printing Office, 1975.

House, E. *Evaluating with validity.* Beverly Hills, Calif.: Sage, 1980.

House, E. R. The logic of evaluative argument. *CSE Monograph Series in Evaluation,* No. 7, 1977.

Jemelka, R. P., and Borich, G. D. Traditional and emerging definitions of educational evaluation. *Evaluation Quarterly,* 1979, *3*(2), 263–276.

Joint Committee on Standards for Educational Evaluation. *Standards for evaluations of educational programs, projects and materials.* New York: McGraw-Hill, 1981.

Kac, M. Some mathematical models in science. *Science,* 1969, *166,* 695–699.

Kaplan, A. *The conduct of inquiry.* San Francisco: Chandler, 1964.

Kaplan, A. Values in inquiry. In G. Riley (Ed.), *Values, objectivity and the social sciences.* Reading, Mass.: Addison-Wesley, 1974.

Kaufman, R. *Educational system planning.* Englewood Cliffs, N.J.: Prentice-Hall, 1972.

Kaufman, R. *Identifying and solving problems: A systems approach.* La Jolla, Calif.: University Associates Publishers, 1976.

Kaufman, R. A possible taxonomy of needs assessments. *Educational Technology,* 1977, *17*(11), 60–64.

Kaufman, R., and English, F. W. *Needs assessment.* Englewood Cliffs, N.J.: Educational Technology Publications, 1979.

Kiresuk, T. J., and Lund, S. H. Goal attainment scaling. In C. C. Attkisson, W. A. Hargreaves, M. J. Horowitz, and J. E. Sorensen (Eds.), *Evaluation of human service programs.* New York: Academic Press, 1978.

Kiresuk, T. J., and Sherman, R. E. Goal attainment scaling: A general method for evaluating comprehensive community mental health programs. *Community Mental Health Journal,* 1968, *4,* 443–453.

Koch, S. Psychology cannot be a coherent science. *Psychology Today,* 1969, *3*(4), 14, 64, 66–68.

Kuhn, T. S. *The structure of scientific revolutions.* Chicago: University of Chicago Press, 1970.

Lasswell, H. D. *The communication of ideas.* New York: Harper and Row, 1948.

Lasswell, H. D. *Psychotherapy and politics.* New York: Viking Press, 1960.

Laszlo, E. *The systems view of the world.* New York: Braziller, 1972.

Levine, M. Scientific method and the adversary model. *American Psychologist,* 1974, *29*(9), 661–677.

Lewin, K. *Field theory in the social sciences.* New York: Harper and Brothers, 1951.

Lilienfeld, R. *The rise of systems theory: An ideological analysis.* New York: John Wiley and Sons, 1978.

Lindblom, C. *The intelligence of democracy: Decision making through mutual adjustment.* New York: Free Press, 1965.

Lindblom, C. *Strategies for decision making.* Urbana: University of Illinois, Department of Political Science, 1972.

Lindblom, C., and Cohen, D. *Usable knowledge.* New Haven, Conn.: Yale University Press, 1979.

Linowes, D. F. Survival recipe: Socio-economic management. *The Urban and Social Change Review,* 1974, *7*(1), 13–16.

Los Angeles Times News Service. Modern technology harmful if used in wrong situation. Austin, Tx: *Austin American Statesman,* July 23, 1978, p. C8.

Louch, A. R. *Explanation and human action.* Berkeley: University of California Press, 1966.

Lund, D. A. Mental health program evaluation: Where do you start? *Journal of Evaluation and Program Planning,* 1978, *1,* 31–40.

McGrath, J. H. Relevance trees. In S. P. Hencley and J. R. Yates (Eds.), *Futurism in education: Methodologies.* Berkeley, Calif.: McCutchan, 1974.

Mann, J. The outcome of evaluative research. In C. H. Weiss (Ed.), *Evaluating action programs: Readings in social action and education.* Boston: Allyn and Bacon, 1972.

Martino, J. P. *Technological forecasting for decision making.* New York: American Elsevier, 1972.

Meadows, D. H., Meadows, D. L., Randers, J., and Behrens, W. W., III. *The limits to growth* (2nd Ed.). New York: Universe Publications, 1972.

Mesarovic, M., Macko, D., and Takahara, Y. *Theory of hierarchical multilevel systems.* New York: Academic Press, 1970.

Metfessel, N. W., and Michael, W. B. A paradigm involving multiple criterion measures for the evaluation of the effectiveness of school programs. *Educational and Psychological Measurement,* 1967, 931–943.

Mitroff, I. I., Emshoff, J. R., and Kilmann, R. H. Assumptional analysis: A methodology for strategic problem solving. In L. E. Datta and R. Perloff (Eds.), *Improving evaluations.* Beverly Hills, Calif.: Sage, 1979.

Mixon, D. Instead of deception. *Journal for the Theory of Social Behavior,* 1972, *2*(2), 145–177.

Mogulof, M. Elements of a special-revenue-sharing proposal for the social service. Goal setting, decategorization, planning and evaluation. *The Social Service Review,* 1973, *47*, 593–604.

Montemerlo, M. D., and Tennyson, M. E. *Instructional systems development: Conceptual analysis and comprehensive bibliography.* (NAVTRAE-QUIPCEN IH-257). Orlando, Fla.: Naval Training Equipment Center, February 1976.

Morell, J. A. *Program evaluation in social research.* New York: Pergamon Press, 1979.

Murphy, J. *Getting the facts: A fieldwork guide for evaluators and policy analysts.* Santa Monica, Calif.: Goodyear, 1980.

Murrell, S. Conducting a program evaluation: Collaboration, feedback, and open-system perspectives. In R. D. Coursey (Ed.), *Program evaluation for mental health: Methods, strategies, participants.* New York: Grune and Stratton, 1977.

Nagel, S., and Neef, M. *Operations research methods.* Beverly Hills, Calif.: Sage Publications, 1977.

Nathan, R. P. Social policy in the 1970s. *The Urban and Social Change Review,* 1974, *7*(1), 3–6.

Neuber, K., Atkins, W., Jacobson, J., and Reuterman, N. *Needs assessment: A model for community planning.* Beverly Hills, Calif.: Sage, 1980.

Osborn, R. N., and Hunt, J. G. Environment and organizational effectiveness. *Administrative Science Quarterly,* 1974, *19*, 231–246.

Owens, T. R. Educational evaluation by adversary proceedings. In E. R. House (Ed.). *School evaluation: The politics and process.* Berkeley, Calif.: McCutchan, 1973.

Pace, C. R., and Friedlander J. Approaches to evaluation: Models and perspectives. In G. Hanson (Ed.), *Evaluating program effectiveness: New directions for student services.* San Francisco: Jossey-Bass, 1978.

Paige, H. How to plan and control with PERT. *The Harvard Business Review,* 1963, *41*, 87–95.

Parlett, M., and Hamilton, D. Evaluation as illumination: A new approach to the study of innovatory programs. In G. V. Glass (Ed.), *Evaluation Studies Review Manual* (Vol. 1). Beverly Hills, Calif.: Sage Publications, 1976.

Patton, M. Q. *Utilization focused evaluation.* Beverly Hills, Calif.: Sage, 1978.

Patton, M. Q. *Qualitative evaluation methods.* Beverly Hills, Calif.: Sage, 1980.

Perkins, D. N. Evaluating social interventions: A conceptual schema. *Evaluation Quarterly,* 1977, *1*(4), 639–656.

Pfieffer, J. *New look at education: Systems analysis in our schools and colleges.* New York: Odyssey Press, 1968.

Provus, M. *Discrepancy evaluation.* Berkeley, Calif.: McCutchan, 1971.

Proxmire, W. *Report from wasteland.* New York: Praeger Publishers, 1970.

Rapoport, A. The promise and pitfalls of information theory. In W. Buckley (Ed.), *Modern systems research for the behavioral scientist: A sourcebook.* Chicago: Aldine, 1968. (a)

Rapoport, A. Foreword. In W. Buckley (Ed.), *Modern systems research for the behavioral scientist.* Chicago: Aldine, 1968. (b)

Reichardt, C. S., and Cook, T. D. Beyond qualitative versus quantitative methods. In T. D. Cook and C. S. Reichardt (Eds.), *Qualitative and quantitative methods in evaluation research.* Beverly Hills, Calif.: Sage Publications, 1979.

Rein, M. *Social science and public policy.* New York: Penguin, 1976.

Riecken, H. W. *The volunteer work camp: A psychological evaluation.* Cambridge, Mass.: Addison-Wesley, 1952.

Riecken, H. W. *Memorandum on program evaluation.* Internal memorandum to the Ford Foundation, October 1953. (Reprinted in C. H. Weiss (Ed.), *Evaluating action programs: Readings in social action and education.* Boston: Allyn and Bacon, 1972.)

Riecken, H. W., Boruch, R. F. (Eds.). *Social experiments: A method for planning and evaluating social programs.* New York: Seminar Press, 1974.

Rippey, R. M. *Studies in transactional evaluation.* Berkeley, Calif.: McCutchan, 1973.

Rist, R. C. Blitzkrieg ethnography: On the transformation of a method into a movement. *Educational Researcher,* 1980, 9(2), 8-10.

Rivlin, S. *Systematic thinking for social action.* Washington, D.C.: Brookings Institute, 1971.

Rochberg, R., Gordon, J. J., and Helmer, O. *The use of cross impact matrices for forecasting and planning.* Middletown, Conn.: Institute for the Future, April 1970.

Ross, D. Structured analysis (SA): A language for communicating ideas. *IEEE Transactions on Software Engineering,* 1977, *SE-3*(1), 16-34.

Ross, D., and Schoman, K. Structured analysis for requirements definition. *IEEE Transactions on Software Engineering,* 1977, *SE-3*(1), 6-15.

Rossi, P. H., Freeman, H. E., and Wright, S. R. *Evaluation: A systematic approach.* Beverly Hills, Calif.: Sage Publications, 1979.

Rossi, P. H., and Williams, W. *Evaluating social programs.* New York: Seminar Press, 1972.

Rucker, W. R., Arnspiger, A. C., and Brodbeck, A. J. *Human values in education.* Dubuque, Iowa: William C. Brown, 1969.

Sage, D. E., and Chobot, R. B. The scenario as an approach to studying the future. In S. P. Hencley and J. R. Yates (Eds.), *Futurism in education: Methodologies.* Berkeley, Calif.: McCutchan, 1974.

Schulberg, H. C., and Baker, F. Program evaluation models and the implementation of research findings. *American Journal of Public Health,* 1968, *58*(7), 1248-1255.

Scriven, M. The methodology of evaluation. *AERA Monograph Series on Curriculum Evaluation, No. 1: Perspective on Curriculum Evaluation.* Chicago: Rand McNally, 1967.

Scriven, M. Goal free evaluation. In E. House (Ed.), *School evaluation.* Berkeley, Calif.: McCutchan, 1973.

Scriven, M. The concept of evaluation. In M. Apple, M. Subkoviak, and H. Lufler, Jr. (Eds.), *Educational evaluation: Analysis and responsibility.* Berkeley, Calif.: McCutchan, 1974.

Sherman, R. Will goal attainment scaling solve the problems of program evaluation in the mental health field? In R. D. Coursey (Ed.), *Program evaluation for mental health: Methods, strategies, participants.* New York: Grune and Stratton, 1977.

Siegel, L. M., Attkisson, C. C., and Carson, L. G. Need identification and program planning in the community context. In C. C. Attkisson, W. A. Hargreaves, M. J. Horowitz, and J. E. Sorensen (Eds.), *Evaluation of human service programs.* New York: Academic Press, 1978.

Silvern, L. C. LOGOS: A system language for flowchart modeling. *Educational Technology,* 1969, *9*(6), 18-23.

Silvern, L. C. *The evolution of systems thinking in education* (2nd ed.). Los Angeles, Calif.: Education and Training Consultants, 1971.

Simon, H. A. *The sciences of the artificial.* Cambridge, Mass.: MIT Press, Massachusetts Institute of Technology, 1969.

Simon, H. A. Applying information technology to organization design. *Public Administration Review,* 1973, *33*, 268-278.

Sjoberg, G. Politics, ethics and evaluation research. In M. Guttentag and E. L. Struening (Eds.), *Handbook of evaluation research* (Vol. 2). Beverly Hills, Calif.: Sage Publications, 1975.

Smith, E., and Tyler, R. *Appraising and recording student progress.* New York: Harper and Row, 1942.

Sommer, R. No not research. I said evaluation! *APA Monitor,* January 1977.

Sorensen, J. E., and Elpers, J. R. Developing information systems for human service organizations. In C. C. Attkisson, W. A. Hargreaves, M. J. Horowitz, and J. E. Sorensen (Eds.), *Evaluation of human service programs.* New York: Academic Press, 1978.

Sorensen, J. E., and Grove, H. D. Using cost-outcome and cost-effectiveness analyses for improved program management and accountability. In C. C. Attkisson, W. A. Hargreaves, M. J. Horowitz, and J. E. Sorensen (Eds.), *Evaluation of human service programs.* New York: Academic Press, 1978.

Spencer, L. The federal approach to service integration. *The Urban and Social Change Review,* 1974, *7*(1), 7-13.

Stake, R. Persuasions, not models. *Educational Evaluation and Policy Analysis,* 1981, *3*(1), 83-84.

Stake, R. E. The countenance of educational evaluation. *Teachers College Record,* 1967, *68,* 523-540.

Stake, R. E. *Evaluating the arts in education.* Columbus, Ohio: Charles Merrill, 1975. (a)

Stake, R. E. *Program evaluation, particularly responsive evaluation (Occasional paper no. 5).* Kalamazoo, Mich.: The Evaluation Center, Western Michigan University, 1975. (b)

Stufflebeam, D. L., Foley, W. J., Gephart, W. J., Guba, E. G., Hammond, H. D., Merriman, H. O., and Provus, M. M. *Educational evaluation and decision making.* Itasca, Ill.: Peacock, 1971.

Suchman, E. A. *Evaluative research: Principles and practice in public service and social action programs.* New York: Russell Sage Foundation, 1967.

Suchman, E. A. Action for what? A critique of evaluative research. In C. H. Weiss (Ed.), *Evaluating action programs: Readings in social action and education.* Boston: Allyn and Bacon, 1972.

Sutherland, J. W. *A general systems philosophy for the social and behavioral sciences.* New York: Braziller, 1973.

Talmadge, H. Evaluation of local school community problems: A transactional evaluation approach. *Journal of Research and Development in Education,* 1975, *8,* 32-41.

Thompson, J. D. *Organizations in action.* New York: McGraw-Hill, 1967.

Thorndike, R. L., and Hagen, E. *Measurement and evaluation in psychology and education* (2nd ed.). New York: Wiley, 1969.

Tosi, H., and Carroll, S. *Management: Contingencies, structure, and process.* Chicago: St. Claire Press, 1976.

Tyler, R. W. General statement on evaluation. *Journal of Educational Research,* 1942, *35,* 492-501.

Tyler, R. W. *Basic principles of curriculum and instruction.* Chicago: University of Chicago Press, 1950.

US Army Training and Doctrine Command. *Interservice procedures for instructional system development* (Pamphlet 350-30). Ft. Monroe, Va.: U.S. Army Training and Doctrine Command, August 1975.

van Gigch, J. P. *Applied general systems theory.* New York: Harper and Row, 1978.

Vickers, G. *Value systems and social process.* Middlesex, England: Penguin, 1970.

von Bertalanffy, L. *General systems theory.* New York: Braziller, 1968.

Warheit, G. J., Bell, R., and Schwab, J. *Planning for change: Needs Assessment Approaches.* Rockville, Md.: National Institutes of Mental Health, 1974.

Weick, K. E. *The social psychology of organizing.* Reading, Mass.: Addison-Wesley, 1969.

Weick, K. E. Loosely coupled systems. Presented at the annual meeting of the American Educational Research Association, Boston, April 1980.

Weinberg, G. M. *An introduction to general systems thinking.* New York: Wiley, 1975.

Weiss, C. H. Evaluating educational and social action programs: A "treeful of owls." In C. H. Weiss (Ed.), *Evaluating action programs: Readings in social action and education.* Boston: Allyn and Bacon, 1972. (a)

Weiss, C. H. The politicization of evaluation research. In C. H. Weiss (Ed.), *Evaluating action programs: Readings in social action and education.* Boston: Allyn and Bacon, 1972. (b)

Weiss, C. H. Utilization of evaluation: Toward comparative study. In C. H. Weiss (Ed.), *Evaluating action programs: Readings in social action and education.* Boston: Allyn and Bacon, 1972. (c)

Weiss, C. H. (Ed.). *Using social research in public policy making.* Lexington, Mass.: D. C. Heath and Co., 1977. (a)

Weiss, C. H. The politics of evaluation. In W. Neigher, R. J. Hammer, and G. Landsburg (Eds.), *Emerging developments in mental health evaluation. (Proceedings of the HEW-NIMH Region II Program Evaluation Conference, May 1976).* New York: Argold Press, 1977. (b)

Weiss, R. S., and Rein M. The evaluation of broad-aim programs: Difficulties in experimental design and an alternative. In C. H. Weiss (Ed.), *Evaluating action programs: Readings in social action and education.* Boston: Allyn and Bacon, 1972.

Wholey, J. S., Nay, J. N., Scanlon, J. W., and Schmidt, R. E. Evaluation: When is it really needed? *Evaluation,* 1975, *2*(2), 89-93.

Wildavsky, A. The self-evaluating organization. *Public Administration Review,* 1972, *32,* 509-520.

Willems, E. P., and Raush, H. L. *Naturalistic viewpoints in psychological research.* New York: Holt, Rinehart and Winston, 1969.

Williams, W., and Elmore, R. F. (Eds.). *The social program implementation.* New York: Academic Press, 1976.

Witkin, B. R. Needs assessment kits, models and tools. *Educational Technology,* 1977, *17,*(11), 5-18.

Wolf, R. L. Trial by jury: A new evaluation method. *Phi Delta Kappan,* 1975, *57,* 185-187.

Wolf, R. L. The use of judicial evaluation methods in the formulation of educational policy. *Educational Evaluation and Policy Analysis,* 1979, *1*(3), 19-28.

Wolf, R. L., and Tymitz, B. Ethnography and reading: Matching inquiry mode to process. *Reading Research Quarterly,* 1976-1977, *12*(1), 5-11.

Worthen, B., and Sanders, J. *Educational evaluation: Theory and practice.* Worthington, Ohio: Charles Jones, 1973.

Zwicky, F. *Discovery, invention, research: Through the morphological approach.* Toronto: MacMillan, 1969.

Author Index

Numbers in italics refer to the pages on which the complete references are listed.

Subject Index